Microgrids/Nanogrids Implementation, Planning, and Operation

Microgrids/Nanogrids Implementation, Planning, and Operation

Editors

Mohamed Benbouzid
S. M. Muyeen

MDPI • Basel • Beijing • Wuhan • Barcelona • Belgrade • Manchester • Tokyo • Cluj • Tianjin

Editors
Mohamed Benbouzid
University of Brest
France

S. M. Muyeen
Qatar University
Qatar

Editorial Office
MDPI
St. Alban-Anlage 66
4052 Basel, Switzerland

This is a reprint of articles from the Special Issue published online in the open access journal *Applied Sciences* (ISSN 2076-3417) (available at: https://www.mdpi.com/journal/applsci/special_issues/Microgrids_Nanogrids).

For citation purposes, cite each article independently as indicated on the article page online and as indicated below:

LastName, A.A.; LastName, B.B.; LastName, C.C. Article Title. *Journal Name* **Year**, *Volume Number*, Page Range.

ISBN 978-3-0365-5651-2 (Hbk)
ISBN 978-3-0365-5652-9 (PDF)

Contents

About the Editors

Mohamed Benbouzid

Mohamed Benbouzid received his B.Sc. degree in electrical engineering from the University of Batna, Batna, Algeria, in 1990; his M.Sc. and Ph.D. degrees in electrical and computer engineering from the National Polytechnic Institute of Grenoble, Grenoble, France, in 1991 and 1994, respectively; and his Habilitation à Diriger des Recherches degree from the University of Amiens, Amiens, France, in 2000.

After receiving his Ph.D., he joined the University of Amiens, where he was an Associate Professor of electrical and computer engineering. Since September 2004, he has been with the University of Brest, Brest, France, where he is a Full Professor of electrical engineering. Prof. Benbouzid is also a Distinguished Professor and a 1000 Talent Expert at the Shanghai Maritime University, Shanghai, China. His main research interests and experience include the analysis, design, and control of electric machines; variable-speed drives for traction, propulsion, and renewable energy applications; and fault diagnosis of electric machines.

Prof. Benbouzid is an IEEE Fellow and a Fellow of the IET. He is the Editor-in-Chief of the *International Journal on Energy Conversion* and the *Applied Sciences* (MDPI) Section on "Electrical, Electronics and Communications Engineering". He is a Subject Editor for the IET *Renewable Power Generation*.

S. M. Muyeen

S. M. Muyeen received his B.Sc. Eng. Degree from Rajshahi University of Engineering and Technology (RUET), Bangladesh, formerly known as Rajshahi Institute of Technology, in 2000 and his M. Eng. And Ph.D. Degrees from Kitami Institute of Technology, Japan, in 2005 and 2008, respectively, all in Electrical and Electronic Engineering.

At present, Prof. Muyeen is working as a full Professor in the Electrical Engineering Department of Qatar University. His research interests are power system stability and control, electrical machines, FACTS, energy storage systems (ESS), renewable energy, and HVDC systems. He has been a keynote and an invited speaker at many international conferences, workshops, and universities. He has published more than 250+ articles in different journals and international conferences. He has published seven books as an author or editor.

Prof. Muyeen is an IEEE Senior Member and Fellow of Engineers Australia. He is serving as Editor/Associate Editor for many prestigious journals from IEEE, IET, and other publishers, including IEEE *Transactions on Energy Conversion*, IEEE *Power Engineering Letters*, IET *Renewable Power Generation*, and IET *Generation, Transmission & Distribution*.

applied sciences

MDPI

Editorial

Special Issue on Microgrids/Nanogrids Implementation, Planning, and Operation

Mohamed Benbouzid [1,*], S. M. Muyeen [2] and Muhammad Fahad Zia [3]

1 Institut de Recherche Dupuy de Lôme (UMR CNRS 6027), University of Brest, 29238 Brest, France
2 Department of Electrical Engineering, Qatar University, Doha 2713, Qatar
3 Department of Electrical and Computer Engineering, American University in Dubai, Dubai 28282, United Arab Emirates
* Correspondence: mohamed.benbouzid@univ-brest.fr

Citation: Benbouzid, M.; Muyeen, S.M.; Zia, M.F. Special Issue on Microgrids/Nanogrids Implementation, Planning, and Operation. *Appl. Sci.* **2022**, *12*, 9916. https://doi.org/10.3390/app12199916

Received: 23 September 2022
Accepted: 27 September 2022
Published: 1 October 2022

1. Introduction

Today's power system faces the challenges of environmental protection, increasing global demand for electricity, high-reliability requirements, clean energy, and planning restrictions. To move towards a green and smart electric power system, centralized generation facilities are being transformed into smaller and more distributed generations. As a result, the microgrid concept is emerging, where a microgrid can operate as a single controllable system and can be viewed as a group of distributed energy loads and resources, which can include many renewable energy sources and energy storage systems. Energy management of a large number of distributed energy resources is required for the reliable operation of the microgrid.

Microgrids can allow better integration of distributed energy storage capacity and renewable energy sources into the power grid, therefore, increasing its efficiency and resilience to natural and man-caused disruptive events. In addition, microgrids and nanogrids are potential solutions for providing a better electrical service for both insufficiently supplied and remote areas. Microgrids networking with optimal energy management will lead to a sort of smart grid with numerous benefits such as reduced cost and enhanced reliability and resiliency [1]. As microgrids, nanogrids are an effective solution to promote renewable energy consumption and build a low-carbon and environmentally friendly power grid. They include small-scale renewable energy harvesters and fixed energy storage units typically installed in commercial and residential buildings.

In this challenging context, the objective of this special session is to address and disseminate state-of-the-art research and development results on the implementation, planning, and operation of microgrids/nanogrids, where energy management is one of the core issues.

2. Special Issue Contributions

This section briefly presents the special issue published papers' main contributions.

Liu et al. in [2] addressed the issues of operation cost and energy waste caused by wind and light abandonment. Indeed, energy management systems efficiency is a major concern for wind-PV and storage electric vehicle systems, where optimized operation and flexible scheduling are coordinated with the power grid. In this context, a time-sharing scheduling strategy has been proposed based on the storage system's state of charge and flexible equipment. A quantum mayfly algorithm has also been designed to implement the strategy. The scheduling results have shown that the proposed time-sharing scheduling strategy can reduce the system cost by 60%, and the method decreases energy waste compared with ordinary scheduling methods when using the quantum mayfly algorithm.

Ahmed et al. in [3], consider power quality problems such as voltage sag, swell, and harmonics can significantly affect the systems performance. To address these issues, the dynamic voltage restorer became very popular while to become effective fast responsive-operating, fast and accurate detection of sag and swell is essential. However, calculating the voltage magnitude, for comparison purposes, can be tricky in the presence of nonlinearities such as harmonics. In this context, a single-phase quasi-type-1 phase-locked loop, including a pre-loop filter composed of a frequency-fixed delayed signal cancellation method, and a two-stage all-pass filter has been proposed. This pre-loop filter is easy to implement and can provide rejection of any measurement offset thanks to the frequency-fixed nature. The carried out experiments have shown that the proposed method is fast in detecting and compensating any grid voltage anomalies to maintain constant load voltage despite voltage sag, swell, and harmonic distortions.

Roldán-Blay et al. in [4], analyzed the benefits of sharing two stand-alone facilities from a reliability standpoint. In this context, a random sequential Monte Carlo simulation was applied to a cooperative microgrid to evaluate the impact of a simple cooperative strategy on cluster reliability. It was found that the reliability of the system increases when cooperation is allowed. Furthermore, at the design stage, this allows for more cost-effective solutions than individual sizing with a similar level of reliability. The proposed study should be beneficial in optimizing energy consumption, generation, and storage in neighboring communities with distributed energy resources.

Nematollahi et al. in [5], addressed the problem of microgrids distributed generations optimal sizing and sitting with the objectives to minimize the resources generation cost and mitigate power losses. In this context, it has been proposed to use the lightning attachment procedure optimization It has been shown that this procedure outperforms the artificial bee colony approach in terms of convergence speed and the accuracy of solution-finding. In addition, to handle PV generation and load forecasting uncertainties, the lognormal distribution model and the Gaussian process quantile regression (GPQR) approaches have been adopted.

Bouzid et al. in [6], addressed the issues of microgrids stability and performance optimization for different types of loads, including the problem of harmonic cancellation and output voltage disturbance mitigation of distributed generation resource systems caused by nonlinear loads. In this context, robust proportional-resonant controllers with a harmonics compensator based on the internal model principle have been proposed. These controllers ensure robust tracking of sinusoidal reference signals in distributed energy resource systems subject to load variation with respect to sinusoidal disturbances.

Kharrich et al. in [7], proposed to address microgrids challenge of optimal design of the hybrid system while considering both economic cost and location installation feasibility. In this context, a smart algorithm has been developed; namely the quasi-oppositional Bonobo optimizer for the optimal economic design of a stand-alone microgrid in Aswan, Egypt. The proposed study achieved results that highlighted the efficiency of the adopted optimization algorithm for different hybrid system scenarios. Furthermore, a sensitivity analysis has shown that the PV system sizing is critical as it significantly impacts the microgrid's overall performance.

Vink et al. in [8], carried out a study to evaluate the potential of long-term forecasting of energy production and demand in microgrids. In this context, in terms of the scale of a research building, both energy demand and production on a long-term scale have been proposed to be predicted using R software in combination with microgrid data records. The proposed study identified supporting evidence supporting the potential to model both solar power generation and the microgrid (building in this study) energy consumption simultaneously, based on the parameter of solar irradiance and in combination with known time parameters affecting building occupancy.

Jackson et al. in [9], proposed a comprehensive overview of state-of-the-art trends in hierarchical microgrid control. This review-based study has shown the variability of coordinated control schemes in terms of objectives and applications. This review highlighted decentralization and centralization technique characteristics, along with non-communication versus communication approaches. Primary and secondary levels were described by showing different modes of operation as a hierarchical control strategy. Furthermore, a comparative study highlighted the pro and cons of controllers, including the inner loop, power sharing, voltage and frequency restoration, and grid synchronization schemes.

3. Conclusions

Although submissions for this special issue have been completed, more in-depth re-search in the field is being collected in a new special issue: "Microgrids/Nanogrids Implementation, Planning, and Operation, 2nd Volume" (https://www.mdpi.com/journal/applsci/special_issues/2Z946DAPHE, accessed on 23 September 2022).

Author Contributions: All authors contributed equally to this editorial. All authors have read and agreed to the published version of the manuscript.

Funding: This research received no external funding.

Acknowledgments: Thanks to all the authors and peer reviewers for their valuable contributions to this special issue. We would also like to express our gratitude to the journal staff for its invaluable support. We hope that you will find this special issue useful and that it will serve as a reference for future work in the field of microgrids implementation, planning, and operation.

Conflicts of Interest: The authors declare no conflict of interest.

References

1. Zia, M.F.; Elbouchikhi, E.; Benbouzid, M. Microgrids Energy Management Systems: A Critical Review on Methods, Solutions, and Prospects. *Appl. Energy* **2018**, *222*, 1033–1055. [CrossRef]
2. Liu, X.; Zhao, M.; Wei, Z.; Lu, M. Economic Optimal Scheduling of Wind–Photovoltaic-Storage with Electric Vehicle Microgrid Based on Quantum Mayfly Algorithm. *Appl. Sci.* **2022**, *12*, 8778. [CrossRef]
3. Ahmed, H.; Biricik, S.; Komurcugil, H.; Benbouzid, M. Enhanced Quasi Type-1 PLL-Based Multi-Functional Control of Single-Phase Dynamic Voltage Restorer. *Appl. Sci.* **2022**, *12*, 146. [CrossRef]
4. Roldán-Blay, C.; Roldán-Porta, C.; Quiles, E.; Escrivá-Escrivá, G. Smart Cooperative Energy Supply Strategy to Increase Reliability in Residential Stand-Alone Photovoltaic Systems. *Appl. Sci.* **2021**, *11*, 11723. [CrossRef]
5. Nematollahi, A.F.; Shahinzadeh, H.; Nafisi, H.; Vahidi, B.; Amirat, Y.; Benbouzid, M. Sizing and Sitting of DERs in Active Distribution Networks Incorporating Load Prevailing Uncertainties Using Probabilistic Approaches. *Appl. Sci.* **2021**, *11*, 4156. [CrossRef]
6. Bouzid, A.E.M.; Zerrougui, M.; Ben Elghali, S.; Beddiar, K.; Benbouzid, M. Robust Resonant Controllers for Distributed Energy Resources in Microgrids. *Appl. Sci.* **2020**, *10*, 8905. [CrossRef]
7. Kharrich, M.; Mohammed, O.H.; Kamel, S.; Selim, A.; Sultan, H.M.; Akherraz, M.; Jurado, F. Development and Implementation of a Novel Optimization Algorithm for Reliable and Economic Grid-Independent Hybrid Power System. *Appl. Sci.* **2020**, *10*, 6604. [CrossRef]
8. Vink, K.; Ankyu, E.; Kikuchi, Y. Long-Term Forecasting Potential of Photo-Voltaic Electricity Generation and Demand Using R. *Appl. Sci.* **2020**, *10*, 4462. [CrossRef]
9. Jackson, R.; Zulkifli, S.A.; Benbouzid, M.; Salimin, S.; Khan, M.H.; Elhassan, G.; Pathan, E. A Comprehensive Motivation of Multilayer Control Levels for Microgrids: Synchronization, Voltage and Frequency Restoration Perspective. *Appl. Sci.* **2020**, *10*, 8355. [CrossRef]

applied sciences

MDPI

Review

A Comprehensive Motivation of Multilayer Control Levels for Microgrids: Synchronization, Voltage and Frequency Restoration Perspective

Ronald Jackson [1], Shamsul Aizam Zulkifli [1,*], Mohamed Benbouzid [2,*], Suriana Salimin [1], Mubashir Hayat Khan [1], Garba Elhassan [1] and Erum Pathan [3]

1 Department of Electrical Engineering, FKEE, Universiti Tun Hussein Onn Malaysia, Johor 86400, Malaysia; he180095@siswa.uthm.edu.my (R.J.); suriana@uthm.edu.my (S.S.); ge180116@siswa.uthm.edu.my (M.H.K.); ge180033@siswa.uthm.edu.my (G.E.)

2 Institut de recherché Dupuy de Lôme (UMR CNRS 6027 IRDL), University of Brest, 29238 Brest, France

3 Department of Electronic Engineering, Quaid-E-Awam University of Engineering, Science & Technology Nawabshah, Sindh 67450, Pakistan; ge130108@siswa.uthm.edu.my

* Correspondence: aizam@uthm.edu.my (S.A.Z.); Mohamed.Benbouzid@univ-brest.fr (M.B.)

Received: 23 October 2020; Accepted: 23 November 2020; Published: 24 November 2020

Abstract: The current paradigm in integrating intermittent renewable energy sources into microgrids presents various technical challenges in terms of reliable operation and control. This paper performs a comprehensive justification of microgrid trends in dominant control strategies. It covers multilayer hierarchical control schemes, which are able to integrate seamlessly with coordinated control strategies. A general overview of the hierarchical control family that includes primary, secondary, tertiary controls is presented. For power sharing accuracy and capability, droop and non-droop-based controllers are comprehensively studied to address further development. The voltage and frequency restoration techniques are discussed thoroughly based on centralized and decentralized method in order to highlights the differences for better comprehend. The comprehensive studies of grid synchronization strategies also overviewed and analyzed under balanced and unbalanced grid conditions. The details studies for each control level are displayed to highlight the benefits and shortcomings of each control method. A future prediction from the authors' point of view is also provided to acknowledge which control is adequate to be adopted in proportion to their products applications and a possibility technique for self-synchronization is given in this paper.

Keywords: microgrid; hierarchical control; primary; secondary control; droop control; frequency restoration; voltage restoration; grid synchronization

1. Introduction

The current paradigm in the renewable green energy sector is on the penetration of renewable energy sources (RES) into the electrical system. RES promises an alternative approach for electricity suppliers to produce more sustainable energy as well as to reduce global warming as reported in decarbonization policies [1]. Due to their abundance, RES can penetrate the existing electrical power system and gradually counter the dependency on conventional fossil fuel generation. Worth mentioning, it is a new challenge to meet the energy demand because the power production is not consistent and the demand increases year by year. Henceforth, power electronic converters have mainly dominated in the implementation to increase the efficiency of RES and the reliability of power production with stable generation. For this reason, an embedded electricity generation at a smaller scale is formed and known as distributed generation (DG) units. With the inconsistency of RES power production, the microgrid (MG) has been proposed to manage power distribution and quality aspects [2]. The fundamental components in microgrids are depicted in Figure 1, including the scenario during islanded and grid-connected modes.

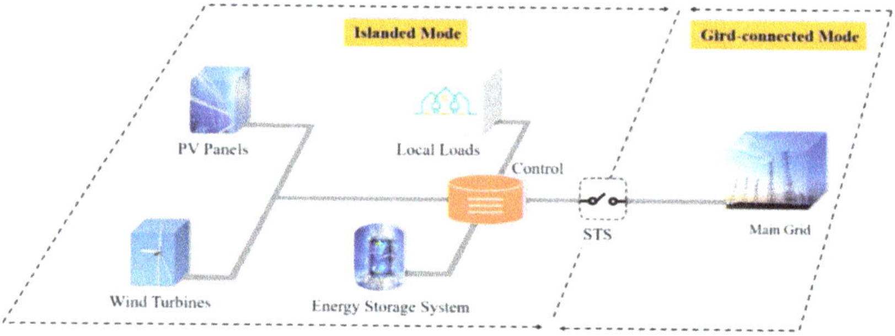

Figure 1. Fundamental microgrid configuration with variable operation modes.

The microgrid can be defined as a cluster of DG units, loads and energy storage system (ESS) operating in coordination to sustainably supply electricity and connecting to the distribution level at a single point of connection, the point of common coupling (PCC). It is designed in such a way so that the power can flow either to the loads and the ESS or to the main grid when the power supply exceeded the microgrid's capacity. A microgrid is capable of operating in two modes: connected to the existing grid network or disconnected from the grid, generally referred to as grid-connected mode (grid-following) and islanded mode (grid-forming), respectively [3–7]. It is achievable because significant power electronic-interfaced converters are acting as voltage source inverters that enables the power converter operates in voltage-controlled mode (VCM) for islanded-mode and current-controlled mode (CCM) during grid-interactive. Henceforth, control strategies also critical impact on power converter interfaced under unbalanced conditions. Theoretically speaking, the MGs controller should be able to determine the accessibility of microgrid operation modes under occurrence of unbalanced loads or weak grids. Thus, the controller mechanism enables the operation of power converter operates either or not accessible in VCM or CCM [8–10]. In VCM, always the inverter yields a controlled output while regulating its amplitude and frequency. On the other hand, CCM adjusts the output current to the reference value and always correlated with the grid synchronization strategies.

The MG's task is to regulate the active and reactive powers generated by the DG units and distribute them according to the load demand. In addition, the surplus power in the microgrid can be exported into the main grid and also can grant essential support to the primary network. In the islanded mode of operation, the microgrid operates as an autonomous entity. Both voltage and frequency are controllable parameters within the DG units, which means that they are no longer supported by the main grid. The active and reactive powers are simultaneously generated within the DG units and accumulated within the microgrid to ensure accurate power load sharing.

There are two definitions of islanding detection: intentional and unintentional islanding. Intentional islanding can be prompted by the scheduled maintenance or when the power quality from the main grid jeopardizes the microgrid operation. Unintentional islanding occurs due to grid faults and unplanned events that are unknown to the microgrid. Hence, an international standard has been established so that microgrids are able to meet several criteria in order to provide reliable ancillary services. The interconnecting of distributed energy resources is stated in IEEE Standards 1547 and 2030 [11–13] so that the transition mode from one state to another can be achieved effectively. For instance, the microgrid can sustain itself when there is a presence of a weak grid and reconnects when the situation is remedied. The assistance of static transfer switch (STS) offers the first action by detecting and disconnecting the microgrid from the grid network, thus guaranteeing continuous power supply to the local loads and protecting the distributed energy source and ESS.

Contributions and Paper Organization

The motivation of this paper is to provide a reviewed on various control strategies of the interconnection of microgrid with the main grid from synchronization, frequency, and voltage restoration perspective under balanced and unbalanced grid conditions. An overview of the microgrid multilayer control trends along with their benefits and shortcomings are carried out in order to help researchers to identify appropriate techniques for their products. There are numerous review studies over the years on this subject matter, however, there is a gap in the discussion on the state-of-the-art coordinated controls in the hierarchy level in proportion of their objectives and applications. A comparative studied for the grid synchronization techniques based on phase-locked loop (PLL) and non-PLL, i.e., synchronverter, frequency-locked loop, discrete Fourier transform, weighted least square estimation are comprehensively discussed and surveyed in detail. The performance comparison of various voltage and frequency regulation controller are also investigated and overviewed in order to provide better power sharing capability. With various multilayer controller strategies and applications, therefore, the findings of this paper should be able to categorized and differentiates the diversification of controller strategies in order to bridge the research gap for further study.

This paper comprises five sections. Microgrid topologies are delivered in Section 2. Section 3 presents the multilayer hierarchical controls with coordinated control. Meanwhile, the primary and secondary state-of-the-art controls are discussed in Sections 3.2 and 3.3, respectively. Section 4 presents key features and recommendations for potential further developments. The conclusion is presented in Section 5.

2. Microgrid Hierarchical Control Group

It is significantly challenging to identify which control method is adequate, and most commonly, the application predominantly depends on the microgrid's traits. The fundamental control objective in a microgrid topology is to ensure reliable and flexible operation, while maintaining close regulation of the microgrid's voltage and frequency as compulsory targets. In general, there are two main state-of-the-art control principles for a microgrid: centralized and decentralized. A centralized control relies on a set of data collection in a dedicated central controller. It continuously monitors and controls the computation taking action at a single central point for the relevant units via low-bandwidth communication between the central controller unit and the controlled groups [14–16]. In contrast with the centralized approach, the decentralized control requires that all DG units and the ESS have a local controller and operate independently based just on local measurements [17–19]. A valid argument between those controls can be attained within the hierarchical control scheme. In this approach, three control levels are distinguished: primary, secondary, and tertiary [20,21], as depicted in Figure 2 with their respective responsibilities. The primary and secondary controllers are responsible for the islanded mode, while the tertiary control mainly serves and is managed at the grid side.

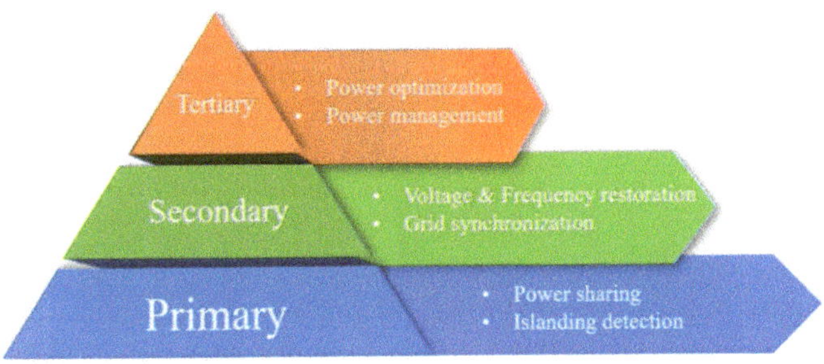

Figure 2. Hierarchical control with respective obligations for each level.

2.1. Primary Control

Also called the internal control or local control, the primary control is defined as the first stage in the control hierarchy group by presenting locally collected information at the fastest response and requires no communication between the controllers. All information are gathered locally and have a certain intelligence level to operate in a decentralized operation mode. According to [22], the decentralized approach should highlight the following fundamental concerns: the integrated units should share the desired total load, stability assurance is provided on a global scope, and the inverter control is capable of providing DC voltage offset avoidance.

The inverter output control's architecture is distinguished from the outer loop for voltage control and internal control for current regulation, as illustrated in Figure 3. The state-of-the-art inner-loop control for the grid-side converter controllers are mainly associated with their reference frames: synchronous frame (*dc*), stationary frame ($\alpha\beta$), and natural frame (*abc*). The synchronous and stationary reference frames served the *dc* and sinusoidal variables, respectively, and are associated with the proportional integral (PI)-based controllers and proportional resonant (PR)-based controllers, respectively. Generally, the natural reference frame is employed in the controllers' approach as PI, PR, dead-beat, and hysteresis controls. The use of PI controllers is the most comprehensive method for control loop design [23].

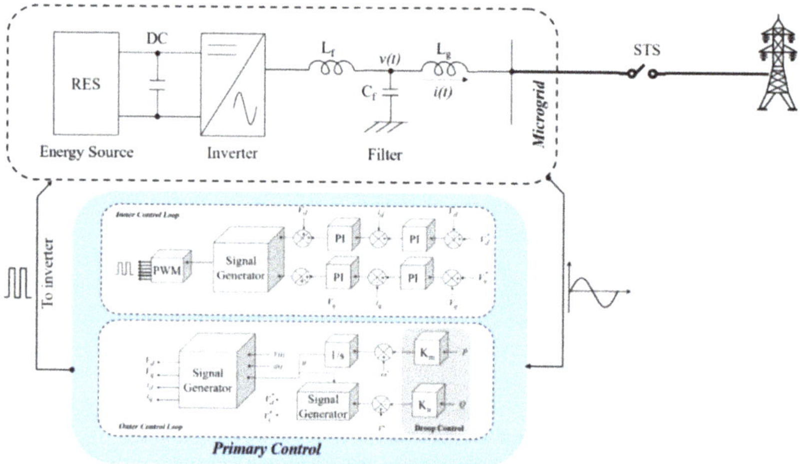

Figure 3. Primary control structure's block diagram.

On the other hand, the state-of-the-art outer loop serves as the power sharing control. This control is significant to control the DG unit as well as the associated local loads to meet the balancing between the active and reactive powers. As for the synchronous generator, three prime components are used to realize the output control and power sharing, which are the voltage regulator, governor, and the inertia obtained from the synchronous machine itself [24]. Droop controllers have been used to emulate the droop traits of the synchronous generator; by purposely simulating the synchronous generator, it has shown the fastest response. It utilizes the correlation of active power-frequency (*P/f*) and reactive power-voltage (*Q/V*) by controlling the values of voltage amplitude and frequency to realize power sharing. The droop-based approach attracts a lot of research efforts due to the advantage it provides over other control options, such as the capability of plug-and-play implementation, flexibility due to non-communication network, simple design, etc. From another perspective, a virtual synchronous generator's (VSG) concept also generally being adopted as presented in [25,26]. The relationship between power productions and voltage-frequency regulations can be achieved through virtual inertia and damping factor into the control loop that mimics the dynamics of the existing synchronous generator. By means the characteristic of the VSG-inverter based is accessible to the main upstream grid system.

In addition, the independency on non-communication control also can be offered as a communication control between the devices and could be applied according to the requirements. Hence, power sharing and voltage regulation could be achieved accurately. Despite adjustment accuracy, communication-link methods can increase the capital cost of microgrids over non-communication methods [27]. Various ways have been presented over the years; for instance, these include master and slave; concentrated control, which has two variants—central limit control and power deviation; instantaneous current sharing; circular chain; and distributed control, among others. Subsequently, the communication-based approach is enormously intricate in extended microgrids and it can degrade system performance during faults at the communication network.

2.2. Secondary Control

The secondary hierarchy control level is subject to reliable monitoring, which guarantees and protects the operation of microgrids in both grid-connected and islanded modes. Monitoring the system variables means supervising both voltage amplitude and frequency deviations to make the required adjustments. It is also referred to as the microgrid energy management system (EMS) by regulating the power flow and power quality within the microgrids. The EMS approach can be classified as centralized and decentralized approaches that serve to manage the unpredictability and intermittent of RESs and load demand [28]. The secondary control guarantees that frequency and voltage deviations are controlled near zero within the load variations or generation in the microgrid and serves the power system by rectifying the grid frequency deviations to within permissible limits [16]. For islanded microgrid operation, the secondary control family is considered as the highest rank in the hierarchical level. According to [2,6], the secondary control has a slower behavior in dynamic response when the secondary unit is detached from primary control, which lessens the communication bandwidth by using sampled measurements of the microgrid variables to permit enough time to execute intricate calculations. A central controller is relatively essential to ensure that the power system operates as seamlessly as possible during significant disturbances and during grid synchronization and transition between the microgrid modes (islanded to grid mode and vice versa) as per market requirements. It can be realized by utilizing an external centralized controller to restore permanent voltage and frequency deviations produced by the primary control.

The architecture of the centralized approach consists of a central controller, which delivers pertinent information of each DG unit and load within the microgrid, and the network itself. The implementation of the centralized method allows for the completion of online optimization routines, given that all the relevant information are collected at a sole point. For instance, the measurements of the main grid's frequency and voltage are realized to generate reference signals and are used by the secondary control loop to provide grid-connected operation mode. The optimal dispatch in [29] mainly performs the analyzed offline calculation in terms of cost and system performance in low-voltage microgrids. In addition, at this level, it is significant to interface the DGs and the ESS via master-slave units [30]. Through this approach, all slave converters can be modeled as current sources. In the islanded operation mode, the battery bank itself operates as a master converter that provides the reference voltage and operating frequency for other sources, which act as slave units. Whereas in grid-connected mode, the latter works as master units and serves to provides the reference voltage and frequency to the PCC. The studies in [31,32] has demonstrated the effectiveness of isolated mode control in an islanded microgrid system including PV system, diesel generator, ESS and even tidal turbine. Other approaches include optimal power flow [33] and terminal sliding mode [34], among others, for several feasible scenarios of the microgrid.

On the other hand, the decentralized method is primarily addressed in the microgrid central controller (MGCC) throughout the multi-agent system (MAS) framework. MAS consists of multiple intelligent agents that operate by relying upon local information, which interacts with one another to achieve global and local objectives. A comprehensive review of the MAS approach is presented in [35]. Meanwhile, MAS implementation is proposed in [36,37]. An approach in [38] includes an additional agent (fuzzy cognitive maps) to reduce the risk of total system breakdown since a possible failure of some decentralized agents is not distinguished from a total EMS failure.

2.3. Tertiary Control

This high rank in the hierarchy control family is primarily associated with the main host grid and generally can be defined as the power flow management according to set points depending on the primary network and is coordinated with the regulations of voltage amplitude and frequency. It is in charge of organizing the operation of multiple microgrids cooperating in the system by measuring the active and reactive power ratio at the PCC, and the comparison of the grid's active and reactive powers against the desired reference values can be realized. The dynamic response of this control is slow in order of several minutes when delivering signals to the secondary level controls at the microgrid and other sub-systems that form the full grid.

According to Figure 4, when the microgrid is operating in grid-tied mode, the power flows can be controlled through frequency adjustment by altering the phase in steady state and the voltage amplitude inside the microgrid [39]. The measured P/Q through the STS, P_G, and Q_G are compared with the desired P^* and Q^* to obtain the desired reference values. Distribution network operator (DNO) and market operator (MO) are among the multivalent controller approaches. DNO is needed when there are more than one microgrids in the distribution system, whereas MO is accountable for the microgrid market management; both are part of the primary grid, which means that they are unaccountable for the microgrid management. Henceforth, there will be no further discussion concerning the tertiary control approach.

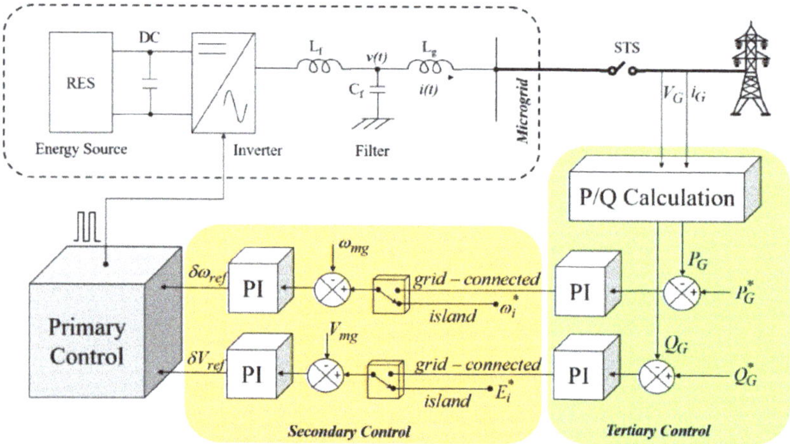

Figure 4. Standard structure's block diagram consisting of secondary and tertiary controls.

3. Hierarchical State-of-the-Art Control

This section discusses the microgrid control family. As previously mentioned, the primary control is responsible for providing the current reference signal according to the voltage source inverter (VSI) operation modes, either in VCM or CCM. These characteristics can be determined according to the electrical grid conditions at the PCC. All the different controller approaches significantly provide better steady-state error requirements and improve the disturbance rejection. A brief discussion on the variability methods from the very simple to the sophisticated analytical approaches is presented here.

However, the primary control permanently causes voltage and frequency variations, which considerably affects the power sharing performance as well as the dynamic response. Other than that, the conventional droop approach does not meet the specific criteria when the parallel system is obligated to share nonlinear load portion because the control units have to consider the harmonic current while, at the same time, retaining the active and reactive powers at the demand level. Henceforth, numerous studies have discussed the secondary control groups to address these inherent issues.

3.1. Overview of the Inner-Loop Control Methods

As previously mentioned, the use of power electronic interfaces is able to make the microgrid operate in both islanded and grid-connected modes. Thus, VSI realizes the flexibility of microgrid operation modes by relying upon VCM and CCM behaviors. For instance, VCM is used in islanded mode to maintain the stability of voltage. In contrast, in grid-connected mode, CCM is commonly used to inject current into the primary grid. The inner current loop is mandatory to provide the current reference associated with the external voltage control loop.

- *Proportional Integral (PI)*: The PI method is the most applied over the years to serve different purposes and is practically employed with the synchronous reference frame associated with *dc* variables. It delivers numerous benefits, such as less complicated controller designs, applicability in linear time-invariant systems, practicability for the single-input-single-output (SISO) system, etc. For instance, [40] discusses the effects of the fractional order of the PI method for load-sharing revamping, and the same method is proposed in [41] and [42] in their respective applications. The PI controller does not solely correspond on the *dc* variables but also on the sinusoidal variables. As mentioned in [43], the PI current controller integrators associate with the stationary frame under undesirable and distorted operating conditions. It offers low computational burden with zero steady-state error for current harmonics concerns.
- *Proportional resonant (PR)*: A viable alternative to the synchronous frame PI controllers is the stationary frame PR compensators [44]. The PR controller is often interpreted as a sinusoidal controller as it is widely associated with the stationary reference frame ($\alpha\beta$) as well as the natural reference frame (*abc*) due to its capability of forcing errors to be zero and for its harmonics compensation [45,46]. Similar methods are proposed in [47] for the distributed power generation systems (DPGS). Studies demonstrate that resonant controllers are an adequate option for inner current control loops to show anticipated performances. It has a high gain around the natural resonant frequency, ω [48], as it is able to introduce a low-order harmonics compensator to improve the current controller. It offers no excessive computational resources, fast transient performance, and the capability of zero steady-state error at the AC frequency, among others.
- *Dead-beat control (DB)*: Among discrete-time linear controls, dead-beat is the most commonly employed due to its fast transient response in different applications, which makes it a suitable approach for current regulation. The DB control is able to place all the closed-loop poles at zero [49]; thus, the tracking error settles to zero within a few sampling steps equivalent to the SISO systems. Worth mentioning, the DB control is in the predictive family, which forecasts the effect of the control action by tracking the current reference accurately without any error. Some in literature have proposed methods to improve the conventional DB controller's limitations, such as being vulnerable to model uncertainties, sensitive to model and parameter mismatch, etc. For instance, [50,51] employ an adaptive self-tuning for the pulse-width modulated (PWM) converter by adopting a delay compensation caused by voltage calculation, synchronous frame rotation, and the PWM converter itself in order to be homogeneously relocated outside the closed-loop control system, thus compensating for their effects on the closed loop's stability. The implementation of the current controller with a higher degree of disturbance rejection enables swift current tracking with higher bandwidth qualities. In [52], the study uses a combination of a DB controller with neural-network identification to present nonlinear estimation problems and to operate as a grid voltage estimator to realize a grid voltage sensorless scheme to guarantee

high-quality power injection. However, while the DB controller provides robustness in its performance, its implementation leads to a suffocated control structure.

- *Hysteresis control (H)*: In the hysteresis approach, the designed controller has to be attached to an adaptive band to achieve fixed switching frequency. The output is the state of the switches that offers variable switching frequency. Worth mentioning, the hysteresis approach is allocated to the predictive control group. Numerous techniques and algorithms to achieve fixed switching frequency are presented in the literature [53–56]. In [57], a hysteresis current regulator is applied for the neutral-point clamped inverter as well as the flying capacitor three-phase inverter, where the measured switched phase-leg voltages are adjusted to the phase leg according to the switching frequency. The hysteresis current regulator is distinguished for its robustness against load variations, for exhibiting high-speed transient performance, and for its simple implementations. Despite the several advantages of the hysteresis controller, it still has drawbacks, such as the high switching frequencies in the inverter due to being interphase-based.
- *Predictive control (P)*: Predictive control is a sophisticated control process satisfying a set of constraints. It uses the system model to forecast the future behavior of the controlled variables. In such a way, it can accurately track the reference current while minimizing the forecasting error within zero error. The model predictive power control (MPPC) and voltage control (MPVC) are proposed in [58]. These schemes can ensure stable DC-bus voltages of BESS as well as a stable AC-bus voltage output and a decent power sharing. Both control schemes provide a simpler control algorithm and better performance. However, the predictive control is prone to network variations. Therefore, in [59], a robust disturbance observer for model predictive control has been proposed for a grid-connected inverter to regulate the current output. The observer is designed to estimate the lumped disturbances (model uncertainties and grid voltage disturbances). An improvement in the reference error tracking is made in [60] by applying an adaptive reference model predictive control to track virtual references rather than actual references. A flexibly modeled virtual multi-input-multi-output (MIMO) generates the error and thus no tuning process is required for different operating points. A seamless transition of mode operations of VSI is presented in [61] through predictive detection and the estimation of control algorithm. It can be achieved through the cost function, thus simplifying the controller's algorithm. The advantage of predictive control design-wise is that it can be implemented more simply depending on the applications. Despite the simple implementation, some predictive controls can be more complicated.
- *Sliding mode control (SMC)*: This control is recognizable by having a robust performance and a fast response time in the variability of system parameters over sizeable operating points [62]. The controller takes vigorous actions if the plant experiences deviations from its average operating point [63]. In such a way, the desired dynamics can be ensured through an essential selection of desired dynamics and control law. However, SMC's performance significantly deteriorates due to the chattering phenomenon in discrete implementation. Therefore, the enhancement of SMC has been made through the output ripple optimization. The works in [24,64] have shown that SMC is employed in the secondary control scheme and is used in both frequency and voltage regulators to enhance power sharing performance and battery charging scheme. SMC shows reliable performance during transients and has better disturbance rejection and low sensitivity to parameter value changes.
- *H-infinity control (H_∞)*: The primary task of the H_∞ controller is to inject a pure sinusoidal current to the main network in both linear and nonlinear loads even during grid voltage distortions. Reducing the disturbance effect and computation delay is vital to avoid instability in the system caused by the digital control delay. Therefore, active damping is presented in [65] by applying a capacitor current feedback into the controller design process. It not only increases the tracking efficiency but also is able to reduce the harmonic noise in current output. In [66], the study distinguishes the performance of reference signal tracking by showing a minimal number of errors. It is capable of adjusting to the system voltage and frequency equivalent to their nominal values

after system load variations. As a result, the H_∞ scheme offers a robust dynamic performance even in model uncertainties and unbalanced conditions, reduces tracking error, and has an easy implementation.

According to the aforementioned, the comprehensive review on various power, current and voltage control strategies are summarized as advantages and shortcomings in Table 1. A microgrid designer needs to identify which potential method is adequate to suit the application. An efficient control application should provide robust dynamic performance.

Table 1. The advantages and shortcomings of the inner-loop controllers' scheme.

Controller	Advantages	Shortcomings
PI [43]	Straightforward implementation. A zero steady-state error in the *dq* reference frame.	Performance degrades under distorted condition. Displays steady-state error in an unbalanced system.
PR [50]	Robust performance during disturbances. No excessive computational required. Relatively zero steady-state error.	Frequency variation sensitivity. Harmonics regulation difficulty.
DB	Accurate current reference tracking within zero error. Closed-loop stability compensation. Fast transient response with lower THD.	Vulnerable to model uncertainties and parameter mismatch. High THD in nonlinear load.
H [55,57]	Fast transient response. A robust operation against load variations. Straightforward implementation.	Resonance problems in high switching frequency. Prone to harmonics issues.
P [67,68]	Obtains a swift transient response. Involves less switching frequency. Provides accurate tracking with low THD.	Vulnerable to parameter variations. High computational burden. A precise filter model is vital.
SMC [62,69]	A robust and dynamic response. Decent performance in adaptivity under parameter changes. Excellent disturbance rejection.	Experiences a chattering phenomenon. Requires excessive computational operation.
H_∞ [70]	Robust performance in both linear and nonlinear loads. Fast response with deficient THD issues. Reduced tracking error.	High computational burden. Relatively slow dynamics.

3.2. *Power Sharing for Primary Control*

Power sharing is allocated in the second phase within the primary control level. This control can be ordered depending on whether or not it utilizes the concept of the droop method. Meanwhile, a power sharing control employing a centralized controller belongs to the secondary control [71]. The advantages and drawbacks of each technique are listed in Table 2.

3.2.1. Droop-Based Control

The droop-based control method is the most widely applied method to control DG units, as distinguished in the literature [3,9,72–77]. The well-established control method is frequency and voltage droop. The main advantage of droop-based control is that it is solely based on local information. Hence, it offers a higher degree of operation in terms of flexibility and reliability. However, the conventional droop control experiences various constraints that could affect the power sharing accuracy among the parallel DG units. These are due to several factors, such as (i) voltage amplitude and frequency deviations, (ii) slow transient response, as it requires low-pass filters (LPF), (iii) vulnerability to nonlinear loads or unbalanced conditions, which introduce circulating currents (harmonic currents), and (iv) the mismatch on the inverter output impedance. Therefore, various research efforts have been presented to counter the inherent limitations of the conventional droop control scheme.

- *Virtual output impedance (VI)*: VI is widely applied to advance the power sharing capability and stability of the system under line mismatch. The inverter output impedance would be in resistive and inductive behaviors. The conventional droop approach in large power systems is entirely operating in inductive line impedance. However, when applied in power electronics, the output impedance relies upon the control strategy used by the inner control loop. In [78], the study demonstrates a robust performance of the inverter with resistive output impedance against numerical errors, disturbances, feeder impedance and parameter mismatch by compensating the voltage drop at the load effect and the droop effect itself. As for inductive impedance [79], adaptive droop control is achieved through a self-adjusted VI to provide the decoupling of active and reactive powers, thus guaranteeing accurate power sharing as well as reliable voltage assistance throughout the voltage compensator. A single line-current feedforward control at the voltage reference is widely applied in the conventional VI approach. An enhancement method is presented in [80] to provide VI at both the fundamental and harmonic frequencies by controlling via the DG current loop and the feedforward PCC voltage, respectively. Thus, VI compensates the impacts of the mismatched in the physical feeder impedances and simultaneously achieves precise adjustment of the DG unit's corresponding impedance at both fundamental and selected harmonic frequencies. An adaptive VI proposed in [81] relies upon an extra small injection of AC signal in the output voltage of the parallel inverter to significantly highlight the unbalanced and harmonic power sharing problems caused by the feeder impedance mismatch. Whereas in [82], an adaptive VI is used for accurate reactive power sharing in an islanded microgrid. The compensation of voltage drop mismatch across the feeders is counteracted by employing communication to facilitate the tuning. In [83], the distributed adaptive VI is employed to suppress large circulating currents caused by the slight differences in both magnitudes and phases in the output voltage of the DG units.

- *Adaptive droop (AD)*: The current paradigm in autonomous power sharing is by utilizing the voltage deviation in multi-terminal DC (MTDC) grids. The regulation of voltage deviation, as well as frequency deviation, can be realized through the outage stage of the voltage source converter (VSC) and instantaneously performs power sharing. This method is becoming an alternative approach for system planners and operators. In MTDC grids, the AD approach is applied to offer a better performance of the VSC, which enables the sharing of the burden of power mismatch. For instance, [84,85] include frequency support and ensure sufficient power sharing by considering the available headroom of each converter station into the control action. This means that when the VSC relatively achieves the operating point (close to limits), it has the capability to constrain the burden of loss sharing to a great extent to the neighboring converters with higher headroom and spare capacity. The relationship of voltage–current–frequency (*V-I-f*) characteristics is derived to establish the correlation between frequency and DC voltage. It can adjust its DC voltage reference autonomously according to grid frequency deviation. Hence, it offers the avoidance of overloading and burden in a desirable way. In [86], it is reported that a better dynamic security assessment is performed and autonomous power sharing is provided following the outage of the VSC. It offers stability-constrained adaptive droop gains through the trajectory sensitivity analysis (TSA)-based approach. Meanwhile, the enhancement of power load sharing and the minimization of circulating currents in low-voltage DC microgrid are presented in [87,88]. It can be achieved by applying AD resistance and instantaneous virtual resistance shifting to compensate the trade-off between the difference in current sharing and the adaptive voltage positioning. Fast transient response is shown through the right-half-plane zero analysis method, where the stability of the system is guaranteed when the series resistance zero is above the equivalent to right-half-plane zero. The integration of wind farm converters in MTDC grids can be found in [89]. Power sharing improvement and voltage deviation minimization under the presence of any faults and power imbalances in the system can be achieved by employing the voltage droop control with the help of the derated operation of the wind farm converters. It offers control freedom over the active power adjustment and provides ancillary services to the primary grid. In [90], the study integrates a wind farm converter with fuzzy logic-based control. The method can update the droop coefficient through the availability of power capacity of the converters, which revamps the conventional fixed droop coefficient. Through this method, the scheme continuously tracks the dynamic behaviors of the converters and realizes desirable responses under different outage scenarios. As found by the authors, the AD scheme ensuring good transient response within proper power sharing, its still experience high computational burden and slow dynamic response.

- *Robust droop (RD)*: An RD controller based on the uncertainty and disturbance estimator (UDE) for the nonaffine nonlinear system is presented in [91–93]. The UDE control algorithm relies upon the assumption that uncertainties and disturbances can be estimated through a filter with an adequate bandwidth. In such a way, the system disturbances and the model nonlinearity can be determined and compensated thoroughly. This method offers a straightforward implementation and does not require inner current control and voltage control schemes. Meanwhile, the regulation of DC-link voltage and robust power sharing in grid-connected VSC-MTDC grids is introduced in [94] by considering the impacts of the droop controller's response, instantaneous power, and DC-side uncertainties. The realization of the controller parameter tuning is through the polynomial method to fix the poles of the equation's trait under possible variations in the droop gain. As a result, it ensures rigid performance under sizeable differences in the converter's operating point to a great extent, as well as under parametric uncertainties and during the occurrence of disturbances. It can increase the system stabilization's effectiveness as well as guaranteeing a robust performance within the system. According to [95], the enhancement of the load voltage drop due to the load and the droop effect can be realized via proportional load sharing. The RD control scheme is integrated with the secondary control. Thus, the load voltage can be retained within the desired range around the nominal value, providing better stability and robustness against disturbances, feeder impedance, and component mismatches. The work in [96] concerns the robust dynamic droop's power sharing for the integration of wind turbines and fuel cells. It employs reverse and direct droop controls, where the reverse droop control is responsible for active and reactive power regulation, while the direct droop control is for the frequency and voltage outputs, which are significantly used for the fuel cells. The adjustment of frequency and voltage within the microgrid can be tracked robustly at specifically designed set points and instantaneously provides better power sharing performance. In the case of a low-inertia wind turbine, a robust control method that is invulnerable against system nonlinearities and changes in the network is reported in [97]. The control scheme exploits multivariable H_∞, which integrates with the centralized multi-input-multi-output (MIMO) controller (set reference of active and reactive powers) and the local measurements of active and reactive powers of the droop controllers. The application has excellent power sharing performance against significant disturbances and interconnection operation. In high-voltage microgrid applications, the work in [98] utilizes the signal detection on the high-voltage side of the coupling transformer as the feedback signal. The line impedance is the dominated part of grid impedance (coupling transformer), and hence the impact of line impedance on the power load sharing can be mitigated significantly, especially on the reactive power relationship.
- *Feedforward control*: An enhanced power flow control is demonstrated in [72] by introducing a feedforward control of the primary grid voltage amplitude and frequency to mitigate the impacts of grid fluctuation on power flow control. It offers improvement in controller stability and provides seamless transitions for grid-connected operation. A similar approach observed in [99] through an adaptive feedforward compensation provides better steady-state power sharing performance while keeping the voltage and frequency regulation. Thus, the robustness of the system stability is enhanced against droop coefficients and network uncertainties.

3.2.2. Non-Droop-Based Control

Apart from droop-based methods, various techniques can be applied that may or may not require communication. These techniques are listed below:

- *Master and slave*: Similarly considered as a secondary control, it consists of a voltage controller for controlling the output voltage to provide a current reference for the other units that assigns the master-slave units. A literature review on this particular method can be found in [30,100–102].

The master unit can be defined through a fixed module arbitrarily or based on the maximum current. In the case of a grid-connected mode, the grid itself is presented as the master unit and no specific control is required for grid-connected and islanding operations. For instance, in [103,104], a supplemental control algorithm is proposed to provide speedy response and sophisticated operating conditions for multiple inverters in an autonomous microgrid; thus, the active and reactive powers are shared accurately according to load demands. In [105], a voltage controller is added with a robust controller that is combined with an automatic master control for the precision of output voltage.

- *Concentrated control*: The operation of this method is based on a central controller by means of a communication link between the central control and the other units. This method consists of the central limit control of power deviation, and their reviews can be found in [106–108]. The current reference value is regulated by the voltage controller. It provides a higher degree of power quality, whether in ideal or transient conditions. However, it requires a high-bandwidth communication link, which increases capital cost.
- *Instantaneous power theory (IPT)*: In [109,110], power sharing could be achieved among parallel DGs devoid of any communication between the devices. An enhancement of IPT is included in [111], which uses a new signal decomposition technique with a great extent of selectivity, thus making it immune to unbalanced and nonlinear power systems. A similar approach in [112] makes the necessary modifications in the fundamental computation to remedy the compensation under any condition of the voltage supply. Another method is presented in [113] for grid-connected VSI under transient conditions, employing an adaptive transformation that instantaneously makes adjustments accordingly to the dynamic voltage conditions, which allows control over constant and oscillating instantaneous power.
- *Direct Power Control (DPC)*: The DPC algorithm has become an alternative for power sharing control, particularly in grid-connected mode of operation. To provide the reference value of power, it solely relies upon the *p-q* theory to adjust the instantaneous active and reactive power errors and maintaining them within a constant hysteresis band [114,115]. Due to the straightforward implementation, it offers several benefits: (1) no grid voltage information is required, as it can be realized through a binary algorithm, (2) compatibility with both sinusoidal and non-sinusoidal voltage signals, and (3) minimal commutation on converter switches.

Following the discussions above, Table 2 addresses the advantages and shortcomings of each control method to deliver an idea that can bridge the gaps among the control approaches. The designed controller should have several features that are immune to the uncertainty model and parametric variations of the microgrid and to load variations. Hence, the advanced algorithms controller complexity can be reduced for further implementation in practical systems.

Table 2. The advantages and shortcomings of the power sharing control.

Technique	Advantages	Shortcomings
Droop-based [79,80]	Straightforward implementation. No critical communication line is required. Significant reliability and flexibility of operations.	Inaccurate power sharing. Performance degrades under nonlinear load and unbalanced conditions. Relatively slow in transient response.
Virtual impedance [83,88]	Fast transient response. Improved current harmonics sharing. Better output voltage with low THD. Excellent voltage and frequency regulation. Better dynamic response.	Relatively poor active and reactive power sharing performance. If the line impedance of the parallel inverters is mismatched, transformation angles will differ and not be synchronized.
Adaptive droop [89,90]	Decent transient response. Good active power sharing. Compensates circulating current. Better voltage and frequency adjustment. Enhances power sharing accuracy and system stability.	Sophisticated implementation. High computational burden. Sluggish dynamic response.
Robust droop [97,101]	Ensures dynamic stability. Better voltage and frequency adjustment. Robust dynamic performance against load variations.	Relatively poor reactive power sharing performance. Involves high THD in current components.
Non-droop-based [106,112,115]	Power sharing enhancement. Straightforward implementation. Improves the dynamic stability of power sharing and quality. Robustness in parametric variation (system and controlled).	Slow dynamic performance. Performance degrades when communication malfunctions at a single point. Communication line increases capital and maintenance cost.

3.3. Voltage and Frequency Restoration for Secondary Control

As well-known, the deviation of voltage and frequency in the primary droop-based controller in a microgrid is permanent. The secondary control is proposed in the hierarchy level to compensate these deviations. Therefore, numerous approaches have been proposed to enhance the research efforts to tackle this issue by considering whether or not to require a central controller.

1. *Centralized controller*: The principle of the centralized control approach is identical with the inner loop control, as described in Section 3. As previously mentioned, MGCC is responsible for power management, voltage and frequency regulations, and the interconnection with the utility grid. The tprincipal operations are via a communication link that receives the information from the primary network when operating in grid-connected mode and passing the references value to the primary control level. The studies in [20,27,66] show examples of this approach. They mainly concern the coordination of the control strategy that depends on the communication network to provide a satisfactory operation and power sharing of generation and storage devices. Worth mentioning, all the decision making, collaboration, and prioritization are achieved more efficiently when information is collected at a single point through the central controller. However, the significant weakness in the MGCC approach is the dependency of the regulations on the central controller, which means that the system performance degrades during the presence of a malfunction at a single point of communications in the central controller. Besides, the communication link among devices offers unreliable operation in terms of maintenance and cost considerations.
2. *Decentralized controller*: A distributed secondary control or MAS network is reportedly a promising method for improving microgrid stability and performance while enhancing the reliability and scalability of the microgrid. The permanent inherent limitations of the primary control, especially the deviations of voltage and frequency, could be counteracted through finite-time and event-triggered control methods, among others.

   - *Consensus algorithm (CA)*: CA is widely employed in voltage and frequency regulation schemes, wherein the reference values are distributed among all the primary controllers. The most popular methods are through a distributed cooperative finite-time secondary control [22,116] by employing a neighbor-based linear consensus algorithm that allows each controller to communicate with each other and also by employing the so-called sparse communication network. While the consensus of the voltage and frequency set reference values is accomplished over an infinite time horizon with exponential convergence, as in [104,117], the authors apply feedback linearization methods for voltage restoration and finite-time control protocol is used to synchronize frequency to the nominal value, simultaneously achieving accurate active power sharing among the DG units. A similar approach is introduced in [118], which is capable of both restoring *V-f* and guaranteeing reactive power sharing as well. The secondary control under switching communication topology is designed in [119], where MAS network is used for controller stability analysis. A distributed approach offers flexibility and reliability in terms of central control avoidance, which means that the failure of a single unit will not degrade the entire system. Meanwhile, in [120], the restoration of the voltage to the nominal value is realized through the dynamic consensus-based method. The enhancement of flexibility and reliability of the microgrid system is ensured without line impedance consideration.

- *Event-triggered*: ETC is another approach for data sharing when a condition is fulfilled or an event is triggered, instead of continuously exchanging data among the DGs unit. In such a way, a sampled data is controlled through a designed mechanism [121]. Numerous methods have been proposed in recent years, such as the time-triggered, event-triggered, and self-triggered sampling methods; their reports can be found in [122–124]. ETC makes a computation that relies upon the measurement error and the last event of the variable states. The error is the difference between the measured and the observed values of the variable states. The benefits of ETC is that it is able to maintain superior closed-loop stability and performance while reducing the number of information exchange among the DGs. A centralized ETC can be found in [125], which provides with an auxiliary controller that corresponds to collect all the variable states. It simplifies the number of controller updates. However, the secondary compensation terms in ETC are realized via a pure integrator that compromise a poor transient response.

The comprehensive review of coordinated controller approaches are summarized as advantages and shortcomings in Table 3. As shown, the restoration of voltage and frequency of the microgrid is a vital issue that needs to be confronted to provide excellent service and better power sharing. Notably, voltage and frequency restorations are always incorporated in secondary level group, yet less studies exposed involvement of primary control.

Table 3. Performance comparison of voltage and frequency of secondary control.

Controller	Advantages	Drawbacks
Centralized [66]	All the decision making, collaboration, and prioritization are achieved more efficiently when information is collected at a single point through the central controller. Provides fast and effective process, while retaining accurate power sharing.	Performance degrades when failure occurs at a single point of communication. Communication links among devices offer unreliable operation in terms of maintenance and cost considerations.
Distributed consensus [116,118]	Increases flexibility and reliability of the microgrid system without taking line impedance consideration. More reliable and has accurate restorations process.	Sophisticated implementation. High computational burden.
Event-triggered [121,123]	Able to maintain superior closed-loop stability and performance. Effectively updates voltage and frequency to nominal values.	High computational burden. Requires a new switch mechanism, which increases malfunction factor.

4. Grid Synchronization for the Secondary Control Unit

The purpose of grid synchronization is to monitor and self-regulate the phase to prevent the deviations of voltage and frequency by minimizing the variance in voltage, frequency, and phase angle between the RES AC generators' output and the grid supply [126,127]. The synchronization unit control loop is used to synchronize the microgrid's phase with the grid. Such synchronization control entails a cluster of all the controllable DG units employed in the secondary control level in the hierarchical control architecture. It provides relatively zero steady-state errors to be translated into the grid-connected operation mode. Grid synchronization is a subjective subtopic that can be classified into primary or secondary control families. However, in this paper, it is considered as a secondary control. As reported in [115], an ideal synchronization must conform to desirable characteristics such as (i) adeptly tracking the primary grid's phase angle, (ii) effectively tracking variations in the frequency and is immune to disturbance signals and high harmonics parameters, and (iii) fast response to the alterations on the primary grid. Numerous grid synchronization methods have been proposed over the years, either based on phase-locked loop (PLL) or non-PLL methods. On the other hand, synchronization without a dedicated synchronization unit has become an alternative for conventional PLL to provide fast and robust operation and simultaneously reduce the computational burden by nonlinear PLL characteristics. Therefore, numerous proposals for self-synchronization techniques have been presented over the years, as reported in [128–134].

Table 4 address the advantage and disadvantages of grid synchronization with significant methods should be considered to present advanced control techniques that are capable of retaining power quality.

Table 4. The advantages and drawbacks of the grid synchronization techniques.

Controller	Advantages	Drawbacks
PLL	Provides accurate synchronization. Better for grid harmonics and noise rejection. Simple implementation.	Vulnerable in high grid faults. In unsymmetrical voltage faults, the second harmonics appear from the negative sequence that propagates through the PLL system and affects the extracted phase angle.
FLL	Immune to phase angle and frequency shifts, harmonics, and uncertainty parameters. Offers reliability under frequency variations, harmonics, and unbalanced voltage.	High computational burden. The overall controller design becomes complex.
DFT	Provides robust phase tracking with fast transient under highly polluted grid. Accurate synchronization even under noise occurrence.	Vulnerable to unbalanced conditions. Phase shift occurs when the sampling is asynchronous with the leading network.
WLSE	Faster and effective in updating frequency and phase angle fault conditions. Simple and easy to implement.	The estimator takes a longer time interval in detecting frequency changes.
S	Reduces the computational burden. Simplifies the overall controller by the withdrawal of the synchronization unit. Provides accurate synchronization process.	Vulnerable to unbalanced and distorted grid voltages. Stability and control parameter tuning is considerably difficult.

4.1. Phase-Locked Loop

Synchronization via PLL is the most widely adopted and is embedded as a dedicated synchronization unit [135–137]. It offers precise tracking of voltage, phase angle, and frequency, and provides simplicity, flexibility, accurate phase angle estimation, etc. However, PLL's performance depreciates when the power converter is integrated into weak grids [138–141]. This scenario is caused by the presence of voltage imbalance and harmonic distortion, double frequency oscillation to the phase error signal, slow dynamic performance, complicated and slow coefficient performance, etc. Numerous modified methods have been presented over the years to mitigate these issues, as shown in [38,134,142–144]. Methods that have been proposed include synchronous reference frame-PLL (SRF-PLL), fixed reference frame PLL (FRF-PLL), enhanced PLL (E-PLL), and second-order generalized integrator PLL (SOGI-PLL), among others.

- SRF-PLL: Studies in [140,145–147] have shown that this technique is the basic scheme in the three-phase system. It offers a rapid and accurate estimation of the phase angle/frequency in ideal grid conditions and offers a straightforward implementation. The operation of SRF-PLL is through the natural frame (*abc*) being transformed into the dq reference frame through Park's transformation. Despite fast and accurate phase estimation in ideal conditions, the performance is highly degraded against abrupt shifting in the phase angle as well as double frequency errors in negative sequence [148], which are caused by frequency deviation, distorted grid voltage, and the presence of harmonics. Several methods with different techniques and algorithms have been proposed to address these issues: decoupled double SRF-PLL (DDSRF-PLL) [147,149], modified SRF-PLL (MSRF-PLL) [150], and adaptive lattice SRF-PLL (ALSRF-PLL) [151].

- SOGI-PLL: SOGI is widely used in the quadrature signal generator (QSG). Also called quadrature-PLL, it seamlessly detects the positive-sequence component to estimate the input signal in-phase and also the quadrature-phase amplitude. Dual SOGI-PLL (DSOGI-PLL) [137,152] corresponds to the transformation of the voltage vectors in the stationary reference frame ($\alpha\beta$) into QSG, which presents the filtered voltage vector. The positive-sequence voltage vector is obtained via the positive-sequence calculator (PFC), and the q-component is equivalent to zero. As a result, it is able to estimate an accurate phase angle under severe fault conditions such as slow dynamics and large frequency overshoots. On the other hand, a frequency feedback loop is adopted for SOGI resonance frequency under a frequency shift. Some results in [153] demonstrate a rather fast dynamic response and harmonics filter capability. A frequency-fixed SOGI-PLL (FFSOGI-PLL) [154,155] has been introduced to enhance the conventional method by eliminating the frequency feedback loop, thereby increasing the speed and accuracy of the frequency trade-off.
- FRF-PLL: FRF-PLL detects the angular frequency and is used for synchronization purposes instead of for phase angle detection. Thus, the fundamental component is provided in fixed-reference-frame coordinates, including both positive and negative sequences, and makes the assumptions of the angular frequency as an uncertain parameter [156,157]. The operation's performance under unbalanced conditions is reported in [142,158] as being robust against variations in angular frequency and the sag or swells in voltage.
- E-PLL: Using E-PLL, the authors in [159] track the three-phase input signal variables in terms of frequency, amplitude, and phase angle. The fundamental operation is through the extraction of the positive-sequence component of the input signal through an adaptive band-pass filter, as it adjusts the transfer function in proportion to the error signal. It shows a higher degree level of transient response and thus delivers smooth and accurate detection of the fundamental parameters within zero steady-state error. As a result, it provides a robust and precise performance under grid contingencies.

4.2. Frequency-Locked Loop

Among recent advancements in grid synchronization for single- and three-phase systems, a new grid synchronization technique has been proposed in [160] called the frequency-locked loop (FLL) technique. This technique is realized in the stationary reference frame and is used to measure the input signal frequency rather than the grid voltage phase. The benefit of this technique is that it does not experience such abrupt phase angle changes. The authors in [161,162] propose the second-order generalized integrator FLL (SOGI-FLL) by implementing two adaptive filters to produce a stationary reference frame that can auto-tune to the grid frequency. This method gives rise to a new structure, called the dual SOGI (DSOGI-FLL), which can estimate the symmetrical components of the grid voltage even under adverse grid conditions. The work in [163] introduces the enhanced FLL (EFLL) for faster dynamic response. This method proposes a generalized design framework of FLL via the Popov hyperstability criterion-based model reference adaptive control. Therefore, a simple structure and convergence speed of the FLL are developed. An adaptive frequency system based on limit cycle oscillator FLL (LCO-FLL) is proposed in [164]. This technique offers a high degree of invulnerability against phase-frequency shifts, harmonics, and voltage sags. Besides, it performs well in a highly polluted grid by ensuring an acceptable transient unto the synchronization process, which achieves the synchrony with the network from any initial condition set. A new multiresonant frequency adaptive synchronization method is proposed in [134,165] that not only can estimate both positive- and negative-sequence components at fundamental frequency but also can detect the following sequence components at harmonic frequencies. This technique uses both harmonious decoupling networks of multiple SOGI and FLL, known as MSOGI-FLL. As a result, the system frequency becomes more adaptive in behavior. The benefit of this method is that the tuning process of different harmonics at the fundamental grid frequency can be deduced. Under highly distorted grid conditions, MSOGI-FLL contributes to detecting accurately with less computation time and low computational burden in symmetrical harmonic components.

4.3. Discrete Fourier Transform (DFT)

The DFT method is considered as the first method of detecting the frequency and harmonics of the periodic signal [166]. Grid synchronization using DFT analysis is a mathematical tool, which transforms a given function from the time domain to the frequency domain. The authors in [133,167] use recursive adaptive window DFT (RDFT) for power converter line synchronization by means of filtering the received grid voltage. This technique offers a high degree of invulnerability against noise. Still, when the DFT time window does not emulate the grid period, a phase shift arises between the filtered voltage and the grid voltage. Hence, an improved phase-detection via sliding DFT (SDFT) has been proposed in [168,169]. This technique displays a robust performance in phase-tracking capability with fast transient response under grid contingencies. This method, which is based on a phase-detection system, is adequate as it requires a lesser amount of operation to obtain a single frequency component. As a result, computational complexity is reduced with a more straightforward implementation than conventional DFT methods. In highly polluted grid conditions, a positive-sequence phase-angle estimation method is presented in [170–172], known as interpolated DFT (IpDFT), which can realize an optimum trade-off between the estimation accuracy and the response time. The proposed technique depends on the obtained sampling rate and can be utilized in the control hierarchy to determine errors and estimate phasors by using analytical analysis, thus generating a one-cycle transient, which is invulnerable to voltage imbalances, harmonics, noises, and grid frequency variations.

4.4. Weighted Least Square Estimation (WLSE)

The synchronization technique of WLSE was introduced back in 2000 [173] and is gaining more attention for its robustness against disturbances that may exist in the power grid. This method is able to estimate separately the positive and negative sequences of frequency and phase angle without significant delay and with covariance resetting, especially under sudden voltage sag or unbalanced conditions. However, to estimate more accurate harmonic phasors, especially under the deviations of power and frequency, a phasor estimator that is realized via the merger of the harmonics components is applied in [174]. It offers a better and more efficient update for the unit reference signal matrix, along with frequency tracking. Thus, the phasor can be maintained as a constant when the power grid frequency deviates from the nominal value. Several works have tried to improve the performance of the sparse algorithm when suffering from the presence of transients on the tracked phase. An extension of the least-square-style WLS is proposed in [175], wherein it allows the measurement of the instantaneous frequency value of real signals for single- or three-phase systems. This technique is applied to the decoupling methods, adaptive filtering of input signals, and frequency estimation. It provides more robustness of system frequency because the estimated frequency is computed using the provided information. A much simpler structure that can be implemented recursively is proposed in [132]. This synchronizer technique compares the sampled grid voltages to complement a straight line to emulate the estimation of the grid phase angle. Hence a frequency filter is added to expand the synchronizer performance, thus making it immune when operating in a polluted grid.

4.5. Synchronverter

A new concept of synchronization by executing the operation of an inverter that is able to simulate the behavior of a synchronous generator (SG) is developed in [129,176], named a synchronverter (S). A synchronverter comprises the mathematical model of the synchronous machine and acts similarly as SG to deliver the voltage supply [177]. Its controller is on the fundamental level of the power controller with an incorporated capacity of voltage and frequency regulation and hence is able to achieve active and reactive power control and voltage and frequency management. An SG is innately ready to synchronize with the power grid; similarly, a synchronverter syncs with the power grid without a synchronization unit because the synchronization function is coordinated in the power controller. This synchronization technique is able to compensate the slow elements present in the closed-loop system consisting of the synchronization unit (PLL), remove the dependency on the inverter controller and the power system, and eliminate the nonlinear factor that influences the accuracy and speed of the synchronization [178]. It broadens the system bandwidth for better analysis and stability, reducing time and increasing the efficiency of synchronization, and considerably enhances the performance of the system, while simultaneously diminishing the complexity of the overall controller. As in [179] the synchronization of the output voltages and frequency is achieved via virtual rotor inertia computations. A modified self-synchronized synchronverter is introduced in [180] for unbalanced conditions in the main grid. This technique can tackle power ripples and balance the grid currents. To achieve these objectives, a resonant controller is used to restrain current harmonics and power ripples. Through the synchronverter scheme, it offers fast response synchronization, accurate phase angle and frequency tracking. Yet the utilization of virtual inertia and damping coefficient might jeopardize the system stability if the selection is inadequate and provide high computational burden.

5. Possibility Implementation of Frequency Self-Restoration Based on *i*-Droop Function and DER Impacts

The penetration and utilization of RES entail the environmentally friendly and economical operation of the whole system. Current research works and technologies deliver numerous solutions and prospects to accomplish these objectives in distinct fields. Meanwhile, project-based studies, developments, and integrations demonstrate the suitability and practicability of different methods. However, there are still research gaps in delivering better performance, particularly for the droop-based approach. The functions of the primary control should be further studied to expand functionalities that do not limit the power sharing scope, thus increasing the reliability and flexibility of the whole system. There is still a shortage of studies and reports on the use of droop control for voltage and frequency regulation without relying on secondary control adjustment. The management of variable energy generation and consumption is still challenging, including actively participating at the consumer side. Figure 5 depicts the frequency and voltage self-restoration function that is presented by the authors for an *i*-droop possibility technique. As noticed, the idea is to simplify the controller's structure so that it can reduce the complexity of the whole system control design and perform accurate power sharing, while retaining voltage and frequency at their nominal values. As for the authors' hypothesis, it should consider an integrated controller that can fulfill both operations of frequency restoration at grid synchronization in autonomous way. Theoretically, the operation solely relies upon the local evaluation, which can achieve self-synchronization. Figure 6 exhibits the prediction of self-frequency restoration under load variations suggested by authors.

The requirements for a simple and reliable operation as well as the autonomous coordination strategies of different distributed energy resources (DERs) are gaining more attention. As known, the selection of centralized or decentralized controller techniques has trade-offs between them. Fully decentralized management is proven to offer flexibility enhancement, but a comprehensive analysis is necessary to provide a secure and reliable operation of the system.

Further penetration of RES is anticipated in microgrid systems as they are distinguished as being pollution-free and thus environment-friendly. Henceforth, additional efforts are essential to solving the challenges in *PQ* issues correlated with RES and grid stability.

Grid synchronization is one of the vital areas in terms of a microgrid's technical challenges, primarily when the microgrid relies upon a dedicated synchronization unit. According to the studies above, there is still a shortage of reports and reviews on the application of artificial intelligence in a microgrid grid-connected synchronization. Significantly, a hybrid method can be initiated with other conventional means. In such a way, self-grid synchronization can be realized at fundamental grid frequencies.

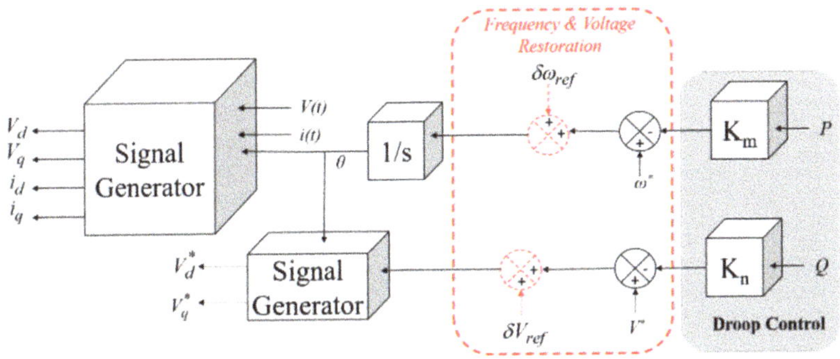

Figure 5. The potential self-restoration function for voltage and frequency in i-droop mode.

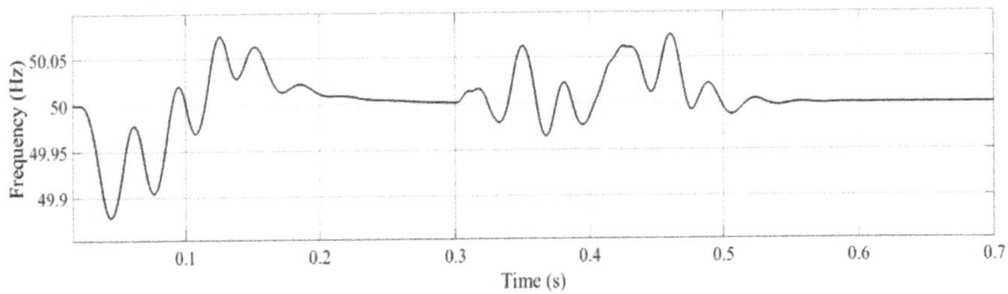

Figure 6. A self-frequency restoration at the fundamental value.

6. Conclusions

An ample summary has been delivered in the overview of the trends in the microgrid's state-of-the-art hierarchical control. The presented review exhibits the variability in the coordinated controller schemes in terms of objectives and applications, which is vital in the development of an intelligent microgrid. This review highlights the features of decentralizing and centralize technique, as well as non-communication against communication approach. The primary and secondary level is described showing different operating modes as a hierarchical control strategy, in terms of the comparative study exhibits the advantages and disadvantages of each controller, particularly the inner-loop, power sharing, voltage and frequency restoration, and grid synchronization schemes.

As for the primary control level, the prime responsibility is to realize power sharing, power quality, flexibility, and reliability. It has been concluded that droop-based method is primarily applied due to its simplicity, plug-and-play feature, and no communication-based requirement while ensuring proper power sharing both in islanded and grid-connected modes of operation. However, it suffers from voltage and frequency deviations, which becomes the main drawback that can jeopardize the microgrid's stability and performance. Therefore, numerous schemes have been proposed to overcome the droop-based method's inherent limitations. For instance, virtual impedances, adaptive control, and robust control are control techniques that can deliver accurate reference signals even in an undesirable state. Secondary level techniques have been identified whether or not implements in a centralized or decentralized approach. According to studies, it has been distinguished that a centralized approach are adequate for single-use in low-scale microgrid configurations. In contrast, the decentralized technique are applicable for multiple users at large-scale microgrid systems, with both voltage-frequency restoration and grid synchronization has been presented. Significantly requires to designs the controller that enables synchronization mechanism, which can embed into the controller loop with no dependency on

PLL unit. Through this study it finds that, there are shortfalls on self-synchronization in decentralized manner that enables voltage and frequency restoration functionality in a manner that resemblance the function of PLL. Henceforth, this concept are recommended for further study.

As aforementioned, the future trends in microgrid technologies are toward advanced decentralized control methods with incorporated an artificial intelligence. The next potential research angles that need to be accomplished for the further evolution of smart microgrids is a straightforward implementation with a robust algorithm in advanced decentralized and self-synchronization techniques suggested by *i*-droop technique that can be a bridge gap for the solution in order to have immunes against undesired conditions. With simple development, less computational involvement and economically wise have becomes an alternative solution for controller designers in order to have fast dynamic response and robustness to high disturbances. Thus, it provides more adequate voltage-frequency self-restoration, fast grid-connection, and yet providing an accurate power sharing that can be seen in next coming papers before the system become fully autonomous.

Funding: This research was funded by UTHM Research Management Centre fund, and Research University Grant Program under grant number H538.

Acknowledgments: The authors would like to give acknowledgement to the Advanced Control in Power Converter research group for essential assistance.

Conflicts of Interest: The authors declare no conflict of interest.

References

1. Meckling, J.; Sterner, T.; Wagner, G. Policy sequencing toward decarbonization. *Nat. Energy* **2017**, *2*, 918–922. [CrossRef]
2. Olivares, D.E.; Mehrizi-Sani, A.; Etemadi, A.H.; Cañizares, C.A.; Iravani, R.; Kazerani, M.; Hajimiragha, A.H.; Gomis-Bellmunt, O.; Saeedifard, M.; Palma-Behnke, R.; et al. Trends in microgrid control. *IEEE Trans. Smart Grid* **2020**, *5*, 1905–1919. [CrossRef]
3. Mueller, J.A.; Kimball, J.W. An Efficient Method of Determining Operating Points of Droop-Controlled Microgrids. *IEEE Trans. Energy Convers.* **2017**, *32*, 1432–1446. [CrossRef]
4. VanDoorn, T.L.; Meersman, B.; De Kooning, J.D.M.; Vandevelde, L. Transition from Islanded to Grid- Connected Mode of Microgrids with Voltage-Based Droop Control. *IEEE Trans. Power Syst.* **2013**, *28*, 2545–2553. [CrossRef]
5. Zhang, D.; Fletcher, J. Operation of Autonomous AC Microgrid at Constant Frequency and with Reactive Power Generation from Grid-forming, Grid-supporting and Grid-feeding Generators. *IEEE Proc. TENCON* **2019**, *10*, 1560–1565.
6. Unamuno, E.; Barrena, J.A. Hybrid ac/dc microgrids—Part II: Review and classification of control strategies. *Renew. Sustain. Energy Rev.* **2015**, *52*, 1123–1134. [CrossRef]
7. Huang, X.; Wang, K.; Qiu, J.; Hang, L.; Li, G.; Wang, X. Decentralized Control of Multi-Parallel Grid-Forming DGs in Islanded Microgrids for Enhanced Transient Performance. *IEEE Access* **2019**, *7*, 17958–17968. [CrossRef]
8. Yoon, S.; Oh, H.; Choi, S. Controller Design and Implementation of Indirect Current Control Based Utility-Interactive Inverter System. *IEEE Trans. Power Electron.* **2013**, *28*, 26–30. [CrossRef]
9. Ketabi, A.; Rajamand, S.S.; Shahidehpour, M. Power sharing in parallel inverters with different types of loads. *IET Gener. Transm. Distrib.* **2017**, *11*, 2438–2447. [CrossRef]
10. Ochs, D.S.; Mirafzal, B.; Sotoodeh, P. A Method of Seamless Transitions Between Grid-Tied and Stand-Alone Modes of Operation for Utility-Interactive Three-Phase Inverters. *IEEE Trans. Ind. Appl.* **2014**, *50*, 1934–1941. [CrossRef]
11. Mariam, L.; Basu, M.; Conlon, M.F. Microgrid: Architecture, policy and future trends. *Renew. Sustain. Energy Rev.* **2016**, *64*, 477–489. [CrossRef]
12. IEEE Standards Coordinating Committee 21. *IEEE Standards 1547 for Interconnecting Distributed Resources with Electric Power Systems*; The Institute of Electrical and Electronic Engineers: New York, NY, USA, 2003.
13. Basso, T. *IEEE 1547 and 2030 Standards for Distributed Energy Resources Interconnection and Interoperability with the Electricity Grid*; National Renewable Energy Laboratory: Golden, CO, USA, 2014.

14. Colet, A.; Alvarez-Cuevas-Figuerola, F.; Ruiz-Alvarez, A.; Gomis-Bellmunt, O.; Sudria-Andreu, A. Centralized and Distributed Active and Reactive Power Control of a Utility Connected Microgrid Using IEC61850. *IEEE Syst. J.* **2011**, *6*, 58–67. [CrossRef]
15. Tsikalakis, A.G.; Hatziargyriou, N.D. Centralized Control for Optimizing Microgrids Operation. *IEEE Transf. Energy Convert.* **2011**, *23*, 1–8.
16. Olivares, D.E.; Canizares, C.A.; Kazerani, M. A centralized optimal energy management system for microgrids. In Proceedings of the 2011 IEEE Power and Energy Society General Meeting, Detroit, MI, USA, 24–28 July 2011; pp. 1–6. [CrossRef]
17. Cai, H.; Xiang, J.; Chen, M.Z.Q.; Wei, W. A decentralized control strategy for photovoltaic sources to unify MPPT and DC-bus voltage regulation. In Proceedings of the 2017 American Control Conference (ACC), Seattle, WA, USA, 24–26 May 2017; Volume 5090, pp. 1–6. [CrossRef]
18. De Matos, J.G.; Silva, F.S.F.; Ribeiro, L.A.D.S. Power Control in AC Isolated Microgrids with Renewable Energy Sources and Energy Storage Systems. *IEEE Trans. Ind. Electron.* **2015**, *62*, 3490–3498.
19. Wu, D.; Tang, F.; Dragicevic, T.; Vasquez, J.C.; Guerrero, J.M. Autonomous Active Power Control for Islanded AC Microgrids with Photovoltaic Generation and Energy Storage System. *IEEE Trans. Energy Convers.* **2014**, *29*, 882–892. [CrossRef]
20. Guerrero, J.M.; Vasquez, J.C.; Matas, J.; De Vicuña, L.G.; Castilla, M. Hierarchical control of droop-controlled AC and DC microgrids—A general approach toward standardization. *IEEE Trans. Ind. Electron.* **2011**, *58*, 158–172. [CrossRef]
21. Meng, L.; Shafiee, Q.; Trecate, G.F.; Karimi, H.; Fulwani, D.; Lu, X.; Guerrero, J.M. Review on Control of DC Microgrids and Multiple Microgrid Clusters. *IEEE J. Emerg. Sel. Top. Power Electron.* **2017**, *5*, 928–948.
22. Guerrero, J.M.; Chandorkar, M.; Lee, T.-L.; Loh, P.C. Advanced Control Architectures for Intelligent Microgrids—Part I: Decentralized and Hierarchical Control. *IEEE Trans. Ind. Electron.* **2013**, *60*, 1254–1262. [CrossRef]
23. Timbus, A.; Liserre, M.; Teodorescu, R.; Rodriguez, P.; Blaabjerg, F. Evaluation of Current Controllers for Distributed Power Generation Systems. *IEEE Trans. Power Electron.* **2009**, *24*, 654–664. [CrossRef]
24. Yan, X.; Cui, Y. Control Method of Parallel Inverters with Self-Synchronizing Characteristics in Distributed Microgrid. *Energies* **2019**, *12*, 3871. [CrossRef]
25. Yao, G.; Lu, Z.; Wang, Y.; Benbouzid, M.; Moreau, L. A Virtual Synchronous Generator Based Hierarchical Control Scheme of Distributed Generation Systems. *Energies* **2017**, *10*, 2049. [CrossRef]
26. El Tawil, T.; Yao, G.; Charpentier, J.F.; Benbouzid, M. Design and analysis of a virtual synchronous generator control strategy in microgrid application for stand-alone sites. *IET Gener. Transm. Distrib.* **2019**, *13*, 2154–2161. [CrossRef]
27. Agundis-Tinajero, G.; Segundo-Ramírez, J.; Visairo-Cruz, N.; Savaghebi, M.; Guerrero, J.M.; Barocio, E. Power flow modeling of islanded AC microgrids with hierarchical control. *Int. J. Electr. Power Energy Syst.* **2019**, *105*, 28–36. [CrossRef]
28. Zia, M.F.; Elbouchikhi, E.; Benbouzid, M. Microgrids energy management systems: A critical review on methods, solutions, and prospects. *Appl. Energy* **2018**, *222*, 1033–1055. [CrossRef]
29. Hernandez-Aramburo, C.; Green, T.; Mugniot, N. Fuel Consumption Minimization of a Microgrid. *IEEE Trans. Ind. Appl.* **2005**, *41*, 673–681. [CrossRef]
30. Almada, J.; Leão, R.P.; Sampaio, R.; Barroso, G.C. A centralized and heuristic approach for energy management of an AC microgrid. *Renew. Sustain. Energy Rev.* **2016**, *60*, 1396–1404. [CrossRef]
31. Belila, A.; Benbouzid, M.; Berkouk, E.; Amirat, Y. On Energy Management Control of a PV-Diesel-ESS Based Microgrid in a Stand-Alone Context. *Energies* **2018**, *11*, 2164. [CrossRef]
32. Zia, M.F.; Elbouchikhi, E.; Benbouzid, M.; Guerrero, J.M. Energy Management System for an Islanded Microgrid With Convex Relaxation. *IEEE Trans. Ind. Appl.* **2019**, *55*, 7175–7185. [CrossRef]
33. Ye, P.; Sun, B.; Yang, B.; Huang, X.; Sun, F. An optimal power flow based algorithm for coordinated secondary voltage control. In Proceedings of the 2009 Asia-Pacific Power and Energy Engineering Conference, Wuhan, China, 27–31 March 2009; pp. 1–4.
34. Shotorbani, A.M.; Ghassem-Zadeh, S.; Mohammadi-Ivatloo, B.; Hosseini, S.H. A distributed secondary scheme with terminal sliding mode controller for energy storages in an islanded microgrid. *Int. J. Electr. Power Energy Syst.* **2017**, *93*, 352–364. [CrossRef]

35. Han, Y.; Zhang, K.; Li, H.; Coelho, E.A.A.; Guerrero, J.M. MAS-Based Distributed Coordinated Control and Optimization in Microgrid and Microgrid Clusters: A Comprehensive Overview. *IEEE Trans. Power Electron.* **2018**, *33*, 6488–6508. [CrossRef]
36. Dimeas, A.L.; Hatziargyriou, N.D. Operation of a Multiagent System for Microgrid Control. *IEEE Trans. Power Syst.* **2005**, *20*, 1447–1455. [CrossRef]
37. Logenthiran, I.; Naayagi, R.T.; Woo, W.L.; Phan, V.T.; Abidi, K. Intelligent Control System for Microgrids Using Multiagent System. *IEEE J. Emerg. Sel. Top. Power Electron.* **2015**, *3*, 1036–1045. [CrossRef]
38. Karavas, C.-S.; Kyriakarakos, G.; Arvanitis, K.G.; Papadakis, G. A multi-agent decentralized energy management system based on distributed intelligence for the design and control of autonomous polygeneration microgrids. *Energy Convers. Manag.* **2015**, *103*, 166–179. [CrossRef]
39. Etxegarai, A.; Eguia, P.; Torres, E.; Iturregi, A.; Valverde, V. Review of grid connection requirements for generation assets in weak power grids. *Renew. Sustain. Energy Rev.* **2015**, *41*, 1501–1514. [CrossRef]
40. Zolfaghari, M.; Arani, A.A.K.; Gharehpetian, G.B.; Abedi, M. A fractional order proportional-integral controller design to improve load sharing between DGs in microgrid. In Proceedings of the 2016 Smart Grids Conference (SGC), Kerman, Iran, 20–21 December 2016; pp. 25–29.
41. Arafat, M.N.; Palle, S.; Sozer, Y.; Husain, I. Transition control strategy between standalone and grid-connected operations of voltage-source inverters. *IEEE Trans. Ind. Appl.* **2012**, *48*, 1516–1525. [CrossRef]
42. Vidal, A.; Yepes, A.G.; Freijedo, F.D.; Lopez, O.; Malvar, J.; Baneira, F.; Doval-Gandoy, J. A Method for Identification of the Equivalent Inductance and Resistance in the Plant Model of Current-Controlled Grid-Tied Converters. *IEEE Trans. Power Electron.* **2015**, *30*, 7245–7261. [CrossRef]
43. Yuan, X.; Allmeling, J.; Merk, W.; Stemmler, H. Stationary frame generalized integrators for current control of active power filters with zero steady state error for current harmonics of concern under unbalanced and distorted operation conditions. *IEEE Trans. Ind. Appl.* **2002**, *38*, 523–532. [CrossRef]
44. Holmes, D.G.; Lipo, T.A.; McGrath, B.P.; Kong, W.Y. Optimized Design of Stationary Frame Three Phase AC Current Regulators. *IEEE Trans. Power Electron.* **2009**, *24*, 2417–2426. [CrossRef]
45. Teodorescu, R.; Blaabjerg, F.; Liserre, M.; Loh, P. Proportional-resonant controllers and filters for grid-connected voltage-source converters. *IEE Proc. Electron. Power Appl.* **2006**, *153*, 750. [CrossRef]
46. Lenwari, W. Optimized design of modified proportional-resonant controller for current control of active filters. In Proceedings of the 2013 IEEE International Conference on Industrial Technology (ICIT), Cape Town, South Africa, 25–28 February 2013; Volume 9, pp. 1–6. [CrossRef]
47. Vidal, A.; Freijedo, F.D.; Yepes, A.G.; Fernandez-Comesana, P.; Malvar, J.; Lopez, Ó.; Doval-Gandoy, J. Assessment and Optimization of the Transient Response of Proportional-Resonant Current Controllers for Distributed Power Generation Systems. *IEEE Trans. Ind. Electron.* **2012**, *60*, 1367–1383. [CrossRef]
48. Zhao, W.; Lu, D.D.C.; Agelidis, V.G. Current control of grid-connected boost inverter with zero steady-state error. *IEEE Trans. Power Electron.* **2011**, *26*, 2825–2834. [CrossRef]
49. Lim, J.S.; Park, C.; Han, J.; Lee, Y.I.; Lim, J.S.; Park, C.; Han, J.; Lee, Y.I. Robust Tracking Control of a Three-Phase DC–AC Inverter for UPS Applications. *IEEE Trans. Ind. Electron.* **2013**, *61*, 4142–4151. [CrossRef]
50. Mohamed, Y.A.-R.I.; El-Saadany, E.F. An Improved Deadbeat Current Control Scheme with a Novel Adaptive Self-Tuning Load Model for a Three-Phase PWM Voltage-Source Inverter. *IEEE Trans. Ind. Electron.* **2007**, *54*, 747–759. [CrossRef]
51. Mohamed, Y.A.-R.I.; Rahman, M.A.; Seethapathy, R. Robust Line-Voltage Sensorless Control and Synchronization of LCL-Filtered Distributed Generation Inverters for High Power Quality Grid Connection. *IEEE Trans. Power Electron.* **2012**, *27*, 87–98. [CrossRef]
52. Mohamed, Y.A.-R.I.; El-Saadany, E.F. Adaptive Discrete-Time Grid-Voltage Sensorless Interfacing Scheme for Grid-Connected DG-Inverters Based on Neural-Network Identification and Deadbeat Current Regulation. *IEEE Trans. Power Electron.* **2008**, *23*, 308–321. [CrossRef]
53. Mohseni, M.; Islam, S.M. A New Vector-Based Hysteresis Current Control Scheme for Three-Phase PWM Voltage-Source Inverters. *IEEE Trans. Power Electron.* **2010**, *25*, 2299–2309. [CrossRef]
54. Rahman, K.M.; Khan, M.R.; Choudhury, M.A.; Rahman, M. Variable-band hysteresis current controllers for PWM voltage-source inverters. *IEEE Trans. Power Electron.* **1997**, *12*, 964–970. [CrossRef]
55. Tilli, A.; Tonielli, A. Sequential design of hysteresis current controller for three-phase inverter. *IEEE Trans. Ind. Electron.* **1998**, *45*, 771–781. [CrossRef]

56. Dey, A.; Rajeevan, P.; Biju, K.; Mathew, K.; Gopakumar, K. A Space-Vector-Based Hysteresis Current Controller for a General n-Level Inverter-Fed Drive with Nearly Constant Switching Frequency Control. *IEEE Trans. Ind. Electron.* **2012**, *60*, 1989–1998. [CrossRef]
57. Davoodnezhad, R.; Holmes, D.G.; McGrath, B. A Novel Three-Level Hysteresis Current Regulation Strategy for Three-Phase Three-Level Inverters. *IEEE Trans. Power Electron.* **2013**, *29*, 6100–6109. [CrossRef]
58. Shan, Y.; Hu, J.; Li, Z.; Guerrero, J.M. A Model Predictive Control for Renewable Energy Based AC Microgrids without Any PID Regulators. *IEEE Trans. Power Electron.* **2018**, *33*, 9122–9126. [CrossRef]
59. Nam, N.N.; Choi, M.; Lee, Y.I. Model Predictive Control of a Grid-Connected Inverter with LCL Filter using Robust Disturbance Observer. *IFAC PapersOnLine* **2019**, *52*, 135–140. [CrossRef]
60. Yang, Y.; Tan, S.-C.; Hui, S.Y.R. Adaptive Reference Model Predictive Control with Improved Performance for Voltage-Source Inverters. *IEEE Trans. Control Syst. Technol.* **2018**, *26*, 724–731. [CrossRef]
61. Li, X.; Zhang, H.; Shadmand, M.B.; Balog, R.S. Model Predictive Control of a Voltage-Source Inverter with Seamless Transition between Islanded and Grid-Connected Operations. *IEEE Trans. Ind. Electron.* **2017**, *64*, 7906–7918. [CrossRef]
62. Abdelmadjid, G.; Seghir, B.M.; Ahmed, S.; Messlem, Y.; Safa, A. Sensorless Sliding Mode Vector Control of Induction Motor Drives. *Int. J. Power Electron. Drive Syst.* **2012**, *2*, 277–284. [CrossRef]
63. Han, Y.; Ma, R.; Cui, J. Adaptive Higher-Order Sliding Mode Control for Islanding and Grid-Connected Operation of a Microgrid. *Energies* **2018**, *11*, 1459. [CrossRef]
64. Safa, A.; Berkouk, E.M.; Messlem, Y.; Gouichiche, A. A robust control algorithm for a multifunctional grid tied inverter to enhance the power quality of a microgrid under unbalanced conditions. *Int. J. Electr. Power Energy Syst.* **2018**, *100*, 253–264. [CrossRef]
65. Jin, W.; Li, Y.; Sun, G.; Bu, L. H∞ Repetitive Control Based on Active Damping with Reduced Computation Delay for LCL-Type Grid-Connected Inverters. *Energies* **2017**, *10*, 586. [CrossRef]
66. Sedhom, B.E.; Hatata, A.Y.; El-Saadawi, M.M.; Abd-Raboh, E.-H.E.; Elkis, B.; Abd-Rabo, H. Robust adaptive H-infinity based controller for islanded microgrid supplying non-linear and unbalanced loads. *IET Smart Grid* **2019**, *2*, 420–435. [CrossRef]
67. Parisio, A.; Rikos, E.; Glielmo, L. A Model Predictive Control Approach to Microgrid Operation Optimization. *IEEE Trans. Control. Syst. Technol.* **2014**, *22*, 1813–1827. [CrossRef]
68. Cortés, P.; Kazmierkowski, M.P.; Kennel, R.M.; Quevedo, D.E.; Rodriguez, J. Predictive control in power electronics and drives. *IEEE Trans. Ind. Electron.* **2008**, *55*, 4312–4324. [CrossRef]
69. Mokhtar, M.; Marei, M.I.; El-Sattar, A.A. An Adaptive Droop Control Scheme for DC Microgrids Integrating Sliding Mode Voltage and Current Controlled Boost Converters. *IEEE Trans. Smart Grid* **2019**, *10*, 1685–1693. [CrossRef]
70. Hornik, T.; Zhong, Q.-C. A Current-Control Strategy for Voltage-Source Inverters in Microgrids Based on H^∞ and Repetitive Control. *IEEE Trans. Power Electron.* **2011**, *26*, 943–952. [CrossRef]
71. Vandoorn, T.L.; Guerrero, J.M.; De Kooning, D.M.; Vasquez, J.C.; Vandevelde, L. Decentralized and centralized control of islanded microgrids including reserve management. *IEEE Ind. Electron. Mag.* **2013**, *7*, 42–55. [CrossRef]
72. Deng, Y.; Tao, Y.; Chen, G.; Li, G.; He, X. Enhanced Power Flow Control for Grid-Connected Droop-Controlled Inverters with Improved Stability. *IEEE Trans. Ind. Electron.* **2016**, *64*, 5919–5929. [CrossRef]
73. Sampaio, L.P.; De Brito, M.A.G.; Alves, M.G.; E Melo, G.A.; Canesin, C.A. Robust control applied to power flow control in single-phase inverter with LCL filter, using droop control and D-stability. In Proceedings of the 2012 10th IEEE/IAS International Conference on Industry Applications, Fortaleza, Brazil, 5–7 November 2012; pp. 1–8.
74. Anand, S.; Fernandes, B.G.; Guerrero, J.M. Distributed Control to Ensure Proportional Load Sharing and Improve Voltage Regulation in Low-Voltage DC Microgrids. *IEEE Trans. Power Electron.* **2013**, *28*, 1900–1913. [CrossRef]
75. Jackson, R.; Zulkifli, S.A.; Sham, N.M.B. Power Flow Control Scheme for Hybrid Single-Phase Energy System Using Droop Control: A Comprehensive Survey. *Int. Rev. Electr. Eng.* **2018**, *13*, 305. [CrossRef]
76. Ding, G.; Gao, F.; Li, R.; Wu, B. A Communicationless PCC Voltage Compensation Using an Improved Droop Control Scheme in Islanding Microgrids. *J. Power Electron.* **2017**, *17*, 294–304. [CrossRef]

77. Shuvra, M.; Chowdhury, B. Autonomous control of smart inverters in grid connected and islanded mode. In Proceedings of the 2017 IEEE Power & Energy Society Innovative Smart Grid Technologies Conference (ISGT), Washington, DC, USA, 23–26 April 2017; pp. 1–5.
78. Li, Y.W.; Kao, C.N. An Accurate Power Control Strategy for Power-Electronics-Interfaced Distributed Generation Units Operating in a Low-Voltage Multibus Microgrid. *IEEE Trans. Power Electron.* **2009**, *24*, 2977–2988. [CrossRef]
79. Lu, Z.; Li, W.; Li, Z.; Chen, X.; Lu, H.H.C.; Dong, N.; Liu, X. Adaptive droop control with self-adjusted virtual impedance for three-phase inverter under unbalanced conditions. In Proceedings of the 2017 IEEE International Symposium on Circuits and Systems (ISCAS), Baltimore, MD, USA, 28–31 May 2017; pp. 6–9. [CrossRef]
80. He, J.; Li, Y.W.; Guerrero, J.M.; Blaabjerg, F.; Vasquez, J.C. An Islanding Microgrid Using Enhanced Virtual Impedance Control Scheme. *IEEE Trans. Power Electron.* **2013**, *28*, 5272–5282. [CrossRef]
81. Liu, B.; Liu, Z.; Liu, J.; An, R.; Zheng, H.; Shi, Y. An Adaptive Virtual Impedance Control Scheme Based on Small-AC-Signal Injection for Unbalanced and Harmonic Power Sharing in Islanded Microgrids. *IEEE Trans. Power Electron.* **2019**, *34*, 12333–12355. [CrossRef]
82. Mahmood, H.; Michaelson, D.; Jiang, J. Accurate Reactive Power Sharing in an Islanded Microgrid Using Adaptive Virtual Impedances. *IEEE Trans. Power Electron.* **2015**, *30*, 1605–1617. [CrossRef]
83. Kim, S.; Han, J.; Sun, Q. Consensus-based improved droop control for suppressing circulating current using adaptive virtual impedance in microgrids. In Proceedings of the 2016 Chinese Control and Decision Conference (CCDC), Yinchuan, China, 28–30 May 2016; pp. 4473–4478.
84. Wang, W.; Li, Y.; Cao, Y.; Haeger, U.; Rehtanz, C. Adaptive Droop Control of VSC-MTDC System for Frequency Support and Power Sharing. *IEEE Trans. Power Syst.* **2017**, *33*, 1264–1274. [CrossRef]
85. Chaudhuri, N.R.; Chaudhuri, B. Adaptive droop control for effective power sharing in multi-terminal DC (MTDC) grids. *IEEE Trans. Power Syst.* **2013**, *28*, 21–29. [CrossRef]
86. Yogarathinam, A.; Chaudhuri, N.R. Stability-Constrained Adaptive Droop for Power Sharing in AC-MTDC Grids. *IEEE Trans. Power Syst.* **2019**, *34*, 1955–1965. [CrossRef]
87. Huang, H.-H.; Hsieh, C.-Y.; Liao, J.-Y.; Chen, K.-H. Adaptive Droop Resistance Technique for Adaptive Voltage Positioning in Boost DC–DC Converters. *IEEE Trans. Power Electron.* **2010**, *26*, 1920–1932. [CrossRef]
88. Augustine, S.; Mishra, M.K.; Lakshminarasamma, N. Adaptive Droop Control Strategy for Load Sharing and Circulating Current Minimization in Standalone DC Low-voltage Microgrid. *IEEE Trans. Energy Convers.* **2015**, *6*, 132–141. [CrossRef]
89. Kumar, A.S.; Padhy, B. An Adaptive Droop Control Strategy forAutonomous Power Sharing and DC Voltage Control in Wind Farm-MTDC grids. *IET Renew. Power Gener.* **2019**, *13*, 3180–3190. [CrossRef]
90. Chen, X.; Wang, L.; Sun, H.; Chen, Y. Fuzzy Logic Based Adaptive Droop Control in Multiterminal HVDC for Wind Power Integration. *IEEE Trans. Energy Convers.* **2017**, *32*, 1200–1208. [CrossRef]
91. Zhong, Q.-C.; Wang, Y.; Ren, B. UDE-Based Robust Droop Control of Inverters in Parallel Operation. *IEEE Trans. Ind. Electron.* **2017**, *64*, 7552–7562. [CrossRef]
92. Wang, Y.; Ren, B.; Zhong, Q.-C. Robust Power Flow Control of Grid-Connected Inverters. *IEEE Trans. Ind. Electron.* **2016**, *63*, 6887–6897. [CrossRef]
93. Ren, B.; Zhong, Q.-C.; Chen, J. Robust Control for a Class of Nonaffine Nonlinear Systems Based on the Uncertainty and Disturbance Estimator. *IEEE Trans. Ind. Electron.* **2015**, *62*, 5881–5888. [CrossRef]
94. Davari, M.; Mohamed, Y.A.-R.I. Robust Droop and DC-Bus Voltage Control for Effective Stabilization and Power Sharing in VSC Multiterminal DC Grids. *IEEE Trans. Power Electron.* **2017**, *33*, 4373–4395. [CrossRef]
95. Zhong, Q.-C. Robust Droop Controller for Accurate Proportional Load Sharing Among Inverters Operated in Parallel. *IEEE Trans. Ind. Electron.* **2013**, *60*, 1281–1290. [CrossRef]
96. Roozbehani, S.; Hagh, M.T.; Zadeh, S.G. Frequency control of islanded wind-powered microgrid based on coordinated robust dynamic droop power sharing. *IET Gener. Transm. Distrib.* **2019**, *13*, 4968–4977. [CrossRef]
97. Hossain, J.; Pota, H.R.; Mahmud, M.A.; Aldeen, M. Robust Control for Power Sharing in Microgrids with Low-Inertia Wind and PV Generators. *IEEE Trans. Sustain. Energy* **2015**, *6*, 1067–1077. [CrossRef]
98. Shuai, Z.; Mo, S.; Wang, J.; Shen, Z.J.; Tian, W.; Feng, Y. Droop control method for load share and voltage regulation in high-voltage microgrids. *J. Mod. Power Syst. Clean Energy* **2016**, *4*, 76–86. [CrossRef]
99. Delghavi, M.B.; Yazdani, A. An Adaptive Feedforward Compensation for Stability Enhancement in Droop-Controlled Inverter-Based Microgrids. *IEEE Trans. Power Deliv.* **2011**, *26*, 1764–1773. [CrossRef]

100. Caldognetto, T.; Tenti, P. Microgrids operation based on master-slave cooperative control. *IEEE J. Emerg. Sel. Top. Power Electron.* **2014**, *2*, 1081–1088. [CrossRef]
101. Meng, X.; Liu, Z.; Liu, J.; Wu, T.; Wang, S.; Liu, B. A seamless transfer strategy based on special master and slave DGs. In Proceedings of the 2017 IEEE 3rd International Future Energy Electronics Conference and ECCE Asia (IFEEC 2017—ECCE Asia), Kaohsiung, Taiwan, 3–7 June 2017; Volume 1, pp. 1553–1558.
102. Cagnano, A.; De Tuglie, E.; Cervi, A.; Stecca, R.; Turri, R.; Vian, A. Re-Synchronization control strategy for master-slave controlled microgrids. In Proceedings of the 2019 1st International Conference on Energy Transition in the Mediterranean Area (SyNERGY MED), Cagliari, Italy, 28–30 May 2019; pp. 1–6.
103. Xu, Y. Robust Finite-Time Control for Autonomous Operation of an Inverter-Based Microgrid. *IEEE Trans. Ind. Inform.* **2017**, *13*, 2717–2725. [CrossRef]
104. Zuo, S.; Davoudi, A.; Song, Y.; Lewis, F.L. Distributed Finite-Time Voltage and Frequency Restoration in Islanded AC Microgrids. *IEEE Trans. Ind. Electron.* **2016**, *63*, 5988–5997. [CrossRef]
105. Chiang, H.-C.; Jen, K.-K.; You, G.-H. Improved droop control method with precise current sharing and voltage regulation. *IET Power Electron.* **2016**, *9*, 789–800. [CrossRef]
106. Ramos, R.; Biel, D.; Guinjoan, F.; Fossas, E. Sliding mode design of distributed central limit control strategy for parallel-connected inverters. In Proceedings of the VIII IEEE International Power Electronics Congress, 2002. Technical Proceedings. CIEP 2002, Guadalajara, Mexico, 24 October 2002; pp. 197–201.
107. Banda, J.; Siri, K. Improved central-limit control for parallel-operation of DC-DC power converters. In Proceedings of the PESC '95—Power Electronics Specialist Conference, Atlanta, GA, USA, 18–22 June 1995; pp. 1104–1110.
108. Weckx, S.; Gonzalez, C.; Driesen, J. Combined Central and Local Active and Reactive Power Control of PV Inverters. *IEEE Trans. Sustain. Energy* **2014**, *5*, 776–784. [CrossRef]
109. Furuhashi, T.; Okuma, S.; Uchikawa, Y. A study on the theory of instantaneous reactive power. *IEEE Trans. Ind. Electron.* **1990**, *37*, 86–90. [CrossRef]
110. Czarnecki, L. Closure on "Instantaneous Reactive Power p-q Theory and Power Properties of Three-Phase Systems". *IEEE Trans. Power Deliv.* **2008**, *23*, 1695–1696. [CrossRef]
111. Harirchi, F.; Simões, M.G. Enhanced Instantaneous Power Theory Decomposition for Power Quality Smart Converter Applications. *IEEE Trans. Power Electron.* **2018**, *33*, 9344–9359. [CrossRef]
112. Herrera, R.; Salmerón, P. Present point of view about the instantaneous reactive power theory. *IET Power Electron.* **2009**, *2*, 484–495. [CrossRef]
113. Montanari, A.A.; Gole, A.M. Enhanced Instantaneous Power Theory for Control of Grid Connected Voltage Sourced Converters under Unbalanced Conditions. *IEEE Trans. Power Electron.* **2016**, *32*, 6652–6660. [CrossRef]
114. Trivedi, T.A.; Jadeja, R.; Bhatt, P. A Review on Direct Power Control for Applications to Grid Connected PWM Converters. *Eng. Technol. Appl. Sci. Res.* **2015**, *5*, 841–849. [CrossRef]
115. Jain, B.; Jain, S.; Nema, R. Control strategies of grid interfaced wind energy conversion system: An overview. *Renew. Sustain. Energy Rev.* **2015**, *47*, 983–996. [CrossRef]
116. Dehkordi, N.M.; Sadati, N.; Hamzeh, M. Distributed Robust Finite-Time Secondary Voltage and Frequency Control of Islanded Microgrids. *IEEE Trans. Power Syst.* **2017**, *32*, 3648–3659. [CrossRef]
117. Guo, F.; Wen, C.; Mao, J.; Song, Y.D. Distributed Secondary Voltage and Frequency Restoration Control of Droop-Controlled Inverter-Based Microgrids. *IEEE Trans. Ind. Electron.* **2015**, *62*, 4355–4364. [CrossRef]
118. Shafiee, Q.; Vasquez, J.C.; Guerrero, J.M. Distributed secondary control for islanded MicroGrids—A networked control systems approach. In Proceedings of the IECON 2012—38th Annual Conference on IEEE Industrial Electronics Society, Montreal, QC, Canada, 25–28 October 2012; pp. 5637–5642. [CrossRef]
119. Wang, X.; Zhang, H.; Li, C. Distributed finite-time cooperative control of droop-controlled microgrids under switching topology. *IET Renew. Power Gener.* **2017**, *11*, 707–714. [CrossRef]
120. Zhou, J.; Kim, S.; Zhang, H.; Sun, Q.; Han, R. Consensus-Based Distributed Control for Accurate Reactive, Harmonic, and Imbalance Power Sharing in Microgrids. *IEEE Trans. Smart Grid* **2018**, *9*, 2453–2467. [CrossRef]
121. Shi, J.; Yue, N.; Weng, S. Distributed event-triggered mechanism for secondary voltage control with microgrids. *Trans. Inst. Meas. Control* **2018**, *41*, 1553–1561. [CrossRef]
122. Abdolmaleki, B.; Shafiee, Q.; Seifi, A.R.; Arefi, M.M.; Blaabjerg, F. A Zeno-Free Event-Triggered Secondary Control for AC Microgrids. *IEEE Trans. Smart Grid* **2020**, *11*, 1905–1916. [CrossRef]

123. Fu, A.; Mazo, M. Traffic Models of Periodic Event-Triggered Control Systems. *IEEE Trans. Autom. Control* **2019**, *64*, 3453–3460. [CrossRef]
124. Heemels, W.P.M.H.; Donkers, M.C.F.; Teel, A.R. Periodic event-triggered control for linear systems. *IEEE Trans. Autom. Control* **2013**, *58*, 847–861. [CrossRef]
125. Qian, T.; Liu, Y.; Zhang, W.; Tang, W.; Shahidehpour, M. Event-Triggered Updating Method in Centralized and Distributed Secondary Controls for Islanded Microgrid Restoration. *IEEE Trans. Smart Grid* **2020**, *11*, 1387–1395. [CrossRef]
126. Yazdani, D.; Pahlevaninezhad, M.; Bakhshai, A. Three-phase grid synchronization techniques for grid connected converters in distributed generation systems. In Proceedings of the 2009 IEEE International Symposium on Industrial Electronics, Seoul, Korea, 5–8 July 2009; pp. 1105–1110. [CrossRef]
127. Cho, C.; Jeon, J.; Kim, J.; Kwon, S.; Park, K.; Kim, S. Active Synchronizing Control of a Microgrid. *IEEE Trans. Power Electron.* **2011**, *26*, 3707–3719. [CrossRef]
128. Zhong, Q.-C.; Ming, W.-L.; Zeng, Y. Self-Synchronized Universal Droop Controller. *IEEE Access* **2016**, *4*, 7145–7153. [CrossRef]
129. Zhong, Q.-C.; Nguyen, P.-L.; Ma, Z.; Sheng, W. Self-Synchronized Synchronverters: Inverters without a Dedicated Synchronization Unit. *IEEE Trans. Power Electron.* **2014**, *29*, 617–630. [CrossRef]
130. Ashabani, M.; Freijedo, F.D.; Golestan, S.; Guerrero, J.M. Inducverters: PLL-Less Converters with Auto-Synchronization and Emulated Inertia Capability. *IEEE Trans. Smart Grid* **2016**, *7*, 1660–1674. [CrossRef]
131. Tang, F.; Guerrero, J.M.; Vasquez, J.C.; Wu, D.; Meng, L. Distributed Active Synchronization Strategy for Microgrid Seamless Reconnection to the Grid under Unbalance and Harmonic Distortion. *IEEE Trans. Smart Grid* **2015**, *6*, 2757–2769. [CrossRef]
132. Ramos, C.J.; Martins, A.P.; Carvalho, A.S. Frequency and Phase-Angle Estimation Using Ordinary Least Squares. *IEEE Trans. Ind. Electron.* **2015**, *62*, 5677–5688. [CrossRef]
133. McGrath, B.P.; Holmes, D.G.; Galloway, J.J.H.G. Power Converter Line Synchronization Using a Discrete Fourier Transform (DFT) Based on a Variable Sample Rate. *IEEE Trans. Power Electron.* **2005**, *20*, 877–884. [CrossRef]
134. Golestan, S.; Guerrero, J.M.; Vasquez, J.C.; Abusorrah, A.M.; Al-Turki, Y. Modeling, Tuning, and Performance Comparison of Second-Order-Generalized-Integrator-Based FLLs. *IEEE Trans. Power Electron.* **2018**, *33*, 10229–10239. [CrossRef]
135. Liu, B.; Zhuo, F.; Zhu, Y.; Yi, H.; Wang, F. A Three-Phase PLL Algorithm Based on Signal Reforming under Distorted Grid Conditions. *IEEE Trans. Power Electron.* **2015**, *30*, 5272–5283. [CrossRef]
136. Yazdani, D.; Bakhshai, A.; Jain, P.K. Grid synchronization techniques for converter interfaced distributed generation systems. In Proceedings of the 2009 IEEE Energy Conversion Congress and Exposition, San Jose, CA, USA, 20–24 September 2009; pp. 2007–2014. [CrossRef]
137. Ali, Z.; Christofides, N.; Hadjidemetriou, L.; Kyriakides, E.; Guo, Q.; Blaabjerg, F. Three-phase phase-locked loop synchronization algorithms for grid-connected renewable energy systems: A review. *Renew. Sustain. Energy Rev.* **2018**, *90*, 434–452. [CrossRef]
138. Rasheduzzaman, M.; Kimball, J.W. Modeling and Tuning of an Improved Delayed-Signal-Cancellation PLL for Microgrid Application. *IEEE Trans. Energy Convers.* **2018**, *34*, 712–721. [CrossRef]
139. Ramezani, M.; Golestan, S.; Li, S.; Guerrero, J.M. A Simple Approach to Enhance the Performance of Complex-Coefficient Filter-Based PLL in Grid-Connected Applications. *IEEE Trans. Ind. Electron.* **2017**, *65*, 5081–5085. [CrossRef]
140. Taul, M.G.; Wang, X.; Davari, P.; Blaabjerg, F. An Overview of Assessment Methods for Synchronization Stability of Grid-Connected Converters under Severe Symmetrical Grid Faults. *IEEE Trans. Power Electron.* **2019**, *34*, 9655–9670. [CrossRef]
141. Reza, S.; Sadeque, F.; Hossain, M.; Ghias, A.; Agelidis, V.G.; Agelidis, V.G. Three-Phase PLL for Grid-Connected Power Converters under Both Amplitude and Phase Unbalanced Conditions. *IEEE Trans. Ind. Electron.* **2019**, *66*, 8881–8891. [CrossRef]
142. Escobar, G.; Martinez-Montejano, M.F.; Valdez, A.A.; Martinez, P.R.; Hernandez-Gomez, M. Fixed-Reference-Frame Phase-Locked Loop for Grid Synchronization under Unbalanced Operation. *IEEE Trans. Ind. Electron.* **2010**, *58*, 1943–1951. [CrossRef]
143. Wu, F.; Zhang, L.; Duan, J. A New Two-Phase Stationary-Frame-Based Enhanced PLL for Three-Phase Grid Synchronization. *IEEE Trans. Circuits Syst. II Express Briefs* **2015**, *62*, 251–255. [CrossRef]

144. Guerrero-Rodríguez, N.F.; Rey-Boué, A.B.; Bueno, E.J.; Ortiz, O.; Reyes-Archundia, E. Synchronization algorithms for grid-connected renewable systems: Overview, tests and comparative analysis. *Renew. Sustain. Energy Rev.* **2017**, *75*, 629–643. [CrossRef]
145. Tong, L.; Zou, X.; Feng, S.; Chen, Y.; Kang, Y.; Huang, Q.; Huang, Y. An SRF-PLL-Based Sensorless Vector Control Using the Predictive Deadbeat Algorithm for the Direct-Driven Permanent Magnet Synchronous Generator. *IEEE Trans. Power Electron.* **2013**, *29*, 2837–2849. [CrossRef]
146. Chung, S.-K. A phase tracking system for three phase utility interface inverters. *IEEE Trans. Power Electron.* **2000**, *15*, 431–438. [CrossRef]
147. Golestan, S.; Monfared, M.; Freijedo, F.D. Design-Oriented Study of Advanced Synchronous Reference Frame Phase-Locked Loops. *IEEE Trans. Power Electron.* **2012**, *28*, 765–778. [CrossRef]
148. Rocabert, J.; Azevedo, G.M.S.; Luna, A.; Guerrero, J.M.; Candela, J.I.; Rodríguez, P. Intelligent Connection Agent for Three-Phase Grid-Connected Microgrids. *IEEE Trans. Power Electron.* **2011**, *26*, 2993–3005. [CrossRef]
149. Rodríguez, P.; Pou, J.; Bergas, J.; Candela, J.I.; Burgos, R.P.; Boroyevich, D. Decoupled double synchronous reference frame PLL for power converters control. *IEEE Trans. Power Electron.* **2007**, *22*, 584–592. [CrossRef]
150. Da Silva, C.H.; Pereira, R.R.; Da Silva, L.E.B.; Lambert-Torres, G.; Bose, B.K.; Ahn, S.U. A Digital PLL Scheme for Three-Phase System Using Modified Synchronous Reference Frame. *IEEE Trans. Ind. Electron.* **2010**, *57*, 3814–3821. [CrossRef]
151. González-Espín, F.J.; Figueres, E.; Garcera, G. An Adaptive Synchronous-Reference-Frame Phase-Locked Loop for Power Quality Improvement in a Polluted Utility Grid. *IEEE Trans. Ind. Electron.* **2012**, *59*, 2718–2731. [CrossRef]
152. Yada, H.K.; Kumar, A.S.; Prakash, K. Improved dual-SOGI control for three-phase unified power quality conditioner under distorted grid and load conditions. In Proceedings of the TENCON 2017—2017 IEEE Region 10 Conference, Penang, Malaysia, 5–8 November 2017; pp. 2536–2541.
153. Golestan, S.; Mousazadeh, S.Y.; Guerrero, J.M.; Vasquez, J.C. A Critical Examination of Frequency-Fixed Second-Order Generalized Integrator-Based Phase-Locked Loops. *IEEE Trans. Power Electron.* **2017**, *32*, 6666–6672. [CrossRef]
154. Xiao, F.; Dong, L.; Li, L.; Liao, X. A Frequency-Fixed SOGI-Based PLL for Single-Phase Grid-Connected Converters. *IEEE Trans. Power Electron.* **2017**, *32*, 1713–1719. [CrossRef]
155. Guan, Q.; Zhang, Y.; Kang, Y.; Guerrero, J.M. Single-Phase Phase-Locked Loop Based on Derivative Elements. *IEEE Trans. Power Electron.* **2016**, *32*, 4411–4420. [CrossRef]
156. Mojiri, M.; Bakhshai, A. Stability analysis of periodic orbit of an adaptive notch filter for frequency estimation of a periodic signal. *Automatica* **2007**, *43*, 450–455. [CrossRef]
157. Yazdani, D.; Mojiri, M.; Bakhshai, A.; Joós, G. A Fast and Accurate Synchronization Technique for Extraction of Symmetrical Components. *IEEE Trans. Power Electron.* **2009**, *24*, 674–684. [CrossRef]
158. Martínez-Montejano, M.; Escobar, G.; Torres-Olguin, R. Fixed reference frame phase-locked loop (FRF-PLL) for unbalanced line voltage conditions. In Proceedings of the 2008 IEEE Power Electronics Specialists Conference, Rhodes, Greece, 15–19 June 2008; pp. 4723–4728.
159. Karimi-Ghartemani, M.; Mojiri, M.; Safaee, A.; Walseth, J.A.; Khajehoddin, S.A.; Jain, P.; Bakhshai, A. A New Phase-Locked Loop System for Three-Phase Applications. *IEEE Trans. Power Electron.* **2013**, *28*, 1208–1218. [CrossRef]
160. Golestan, S.; Guerrero, J.M.; Musavi, F.; Vasquez, J.C. Single-Phase Frequency-Locked Loops: A Comprehensive Review. *IEEE Trans. Power Electron.* **2019**, *34*, 11791–11812. [CrossRef]
161. Rodríguez, P.; Luna, A.; Muñoz-Aguilar, R.S.; Etxeberria-Otadui, I.; Teodorescu, R.; Blaabjerg, F. A Stationary Reference Frame Grid Synchronization System for Three-Phase Grid-Connected Power Converters under Adverse Grid Conditions. *IEEE Trans. Power Electron.* **2012**, *27*, 99–112. [CrossRef]
162. Yi, H.; Wang, X.; Blaabjerg, F.; Zhuo, F. Impedance Analysis of SOGI-FLL-Based Grid Synchronization. *IEEE Trans. Power Electron.* **2017**, *32*, 7409–7413. [CrossRef]
163. He, X.; Geng, H.; Yang, G. A Generalized Design Framework of Notch Filter Based Frequency-Locked Loop for Three-Phase Grid Voltage. *IEEE Trans. Ind. Electron.* **2018**, *65*, 7072–7084. [CrossRef]
164. Vazquez-Oviedo, E.I.; Vázquez, N.; Femat, R. Synchronization Technique of Grid-Connected Power Converters Based on a Limit Cycle Oscillator. *IEEE Trans. Ind. Electron.* **2017**, *65*, 709–717. [CrossRef]

165. Rodriguez, P.; Luna, A.; Candela, I.; Mujal, R.; Teodorescu, R.; Blaabjerg, F. Multiresonant Frequency-Locked Loop for Grid Synchronization of Power Converters Under Distorted Grid Conditions. *IEEE Trans. Ind. Electron.* **2011**, *58*, 127–138. [CrossRef]
166. Overney, F.; Mortara, A. Synchronization of sampling-based measuring systesm. *IEEE Trans. Instrum. Meas.* **2014**, *63*, 89–95. [CrossRef]
167. Reza, S.; Ciobotaru, M.; Agelidis, V.G. Single-Phase Grid Voltage Frequency Estimation Using Teager Energy Operator-Based Technique. *IEEE J. Emerg. Sel. Top. Power Electron.* **2015**, *3*, 1218–1227. [CrossRef]
168. Babu, B.C.; Sridharan, K.; Rosolowski, E.; Leonowicz, Z.M. Analysis of SDFT based phase detection system for grid synchronization of distributed generation systems. *Eng. Sci. Technol. Int. J.* **2014**, *17*, 270–278. [CrossRef]
169. Retegui, R.G.; Gonzalez, S.A.; Funes, M.A.; Maestri, S. Implementation of a novel synchronization method using Sliding Goertzel DFT. In Proceedings of the 2007 IEEE International Symposium on Intelligent Signal Processing, Alcala de Henares, Spain, 3–5 October 2007. [CrossRef]
170. Baradarani, F.; Zadeh, M.R.D.; Zamani, M.A. A Phase-Angle Estimation Method for Synchronization of Grid-Connected Power-Electronic Converters. *IEEE Trans. Power Deliv.* **2014**, *30*, 827–835. [CrossRef]
171. Romano, P.; Paolone, M. Enhanced interpolated-DFT for synchrophasor estimation in FPGAs: Theory, implementation, and validation of a PMU prototype. *IEEE Trans. Instrum. Meas.* **2014**, *63*, 2824–2836. [CrossRef]
172. Belega, D.; Macii, D.; Petri, D. Fast Synchrophasor Estimation by Means of Frequency-Domain and Time-Domain Algorithms. *IEEE Trans. Instrum. Meas.* **2013**, *63*, 388–401. [CrossRef]
173. Song, H.-S.; Nam, K. Instantaneous phase-angle estimation algorithm under unbalanced voltage-sag conditions. *IEE Proc. Gener. Transm. Distrib.* **2000**, *147*, 409–415. [CrossRef]
174. Chen, C.-I. A Phasor Estimator for Synchronization Between Power Grid and Distributed Generation System. *IEEE Trans. Ind. Electron.* **2013**, *60*, 3248–3255. [CrossRef]
175. Kusljevic, M.D.; Tomic, J.J.; Jovanovic, L.D. Frequency Estimation of Three-Phase Power System Using Weighted-Least-Square Algorithm and Adaptive FIR Filtering. *IEEE Trans. Instrum. Meas.* **2009**, *59*, 322–329. [CrossRef]
176. Zhong, Q.-C.; Weiss, G. Synchronverters: Inverters That Mimic Synchronous Generators. *IEEE Trans. Ind. Electron.* **2011**, *58*, 1259–1267. [CrossRef]
177. Bevrani, H.; Ise, T.; Miura, Y. Virtual synchronous generators: A survey and new perspectives. *Int. J. Electr. Power Energy Syst.* **2014**, *54*, 244–254. [CrossRef]
178. Belila, A.; Amirat, Y.; Benbouzid, M.; Berkouk, E.M.; Yao, G. Virtual synchronous generators for voltage synchronization of a hybrid PV-diesel power system. *Int. J. Electr. Power Energy Syst.* **2020**, *117*, 105677. [CrossRef]
179. Wang, X.; Chen, L.; Sun, D.; Zhang, L.; Nian, H. A Modified Self-Synchronized Synchronverter in Unbalanced Power Grids with Balanced Currents and Restrained Power Ripples. *Energies* **2019**, *12*, 923. [CrossRef]
180. Blaabjerg, F.; Teodorescu, R.; Liserre, M.; Timbus, A.V. Overview of Control and Grid Synchronization for Distributed Power Generation Systems. *IEEE Trans. Ind. Electron.* **2006**, *53*, 1398–1409. [CrossRef]

applied sciences

MDPI

Article

Robust Resonant Controllers for Distributed Energy Resources in Microgrids

Allal El Moubarek Bouzid [1], Mohamed Zerrougui [2], Seifeddine Ben Elghali [2], Karim Beddiar [3,*] and Mohamed Benbouzid [4,5,*]

1 Laboratoire des Sciences du Numérique de Nantes (LS2N-UMR CNRS 6004), Ecole Centrale de Nantes, 44321 Nantes, France; Allal-El-Moubarek.Bouzid@ec-nantes.fr

2 Laboratory of Information and Systems (LIS-UMR CNRS 7020), Aix-Marseille University, 13007 Marseille, France; mohamed.zerrougui@lis-lab.fr (M.Z.); seifeddine.benelghali@lis-lab.fr (S.B.E.)

3 Laboratoire d'Innovation Numérique pour les Entreprises et les Apprentissages au service de la Compétitivité des Territoires, Centre des Études Supérieures Industrielles, 2 Avenue de Provence, 29200 Brest, France

4 Institut de Recherche Dupuy de Lôme (UMR CNRS 6027 IRDL), University of Brest, 29238 Brest, France

5 Engineering Logistics College, Shanghai Maritime University, Shanghai 201306, China

* Correspondence: kbeddiar@cesi.fr (K.B.); Mohamed.Benbouzid@univ-brest.fr (M.B.)

Received: 6 November 2020; Accepted: 7 December 2020; Published: 14 December 2020

Abstract: Motivated by the problem of different types and variations of load in micro-grids, this paper presents robust proportional-resonant controllers with a harmonics compensator based on the internal model principle. These controllers ensure robust tracking of sinusoidal reference signals in distributed energy resource systems subject to load variation with respect to sinusoidal disturbances. The distributed generation resource and the resonant controllers are described using the augmented state system approach, allowing the application of the state feedback technique. In order to minimize the tracking error and ensure robustness against perturbation, a set of linear matrix inequalities (LMIs) are addressed for the synthesizing of controller gains. Finally, results obtained in the simulation for resonant compensators with the distributed energy system are presented, in which the controller is applied to the CC-CA inverter.

Keywords: proportional resonant controller; robust control; distributed generation, microgrid; LMI

1. Introduction

Distributed energy resource (DER) systems are extensively used in micro-grids for the connection of renewable energy sources and storage [1]. However, this system must be able to maintain in its output voltage the same characteristics of the voltage provided by the electrical network to which they are connected, i.e., constant frequency and voltage amplitude and sine waveform, even when the primary network presents any failure (distortion, or even cut in the blackout supply). In practice, it is sought that, in the event of a power outage, for example, the equipment connected to the DER output will continue to operate normally to feed the locally connected loads. A basic method to obtain the proposed results is to compare the DER output voltage with a sinusoidal reference signal of the same waveform than the expected signal of the primary electrical network in normal operation [2]. However, the tracking of the reference signal becomes complex due to the non-linearities induced by the association of the different types of loads and the LCL filter connected at the output of the inverter. For example, the wide majority of electronic devices require direct current sources, equipped with a rectifier stage with full diodes bridge, which represents a great source of harmonic distortions due to the current waveform, and consequently the voltage in the grid [3].

Due to the critical nature of the loads, DER performance is regulated by international (ANSI/IEEE, 1986; IEC, 2011) standards [4], imposing conditions for the system's transient and steady state performance. During transient, standards require a small variation in the output voltage amplitude and a fast recovery time when a load is added or removed from the system. In steady state, the output voltage of a DER must be a sinusoidal signal with constant amplitude and frequency. To be considered as a sinusoidal signal when subjected to periodic disturbances caused by non-linear loads, the output voltage THD (Total Harmonic Distortion) and IHD (Individual Harmonic) must be within limits defined in these standards. A fundamental performance requirement of a distributed energy resources (DER) system with sinusoidal output is to provide a voltage with low harmonic distortion, which operates even under uncertainties, parametric variations and non-linearities induced by phenomena such as delay and saturation and disturbances, which are very common in practice. This objective can be fulfilled by choosing an adequate control law.

Recently, considerable literature has been produced around the theme of DER control in microgrids to solve the above-cited problems. A major solution consists in the use of resonant compensators, adjusted to act on the fundamental frequency of the output and most significant harmonics. This method is based on the addition of harmonic modes (sine or cosine functions) in the direct control loop. The ideal resonant controller presents infinite gain at the resonance frequency, which allows us to reject the other components. The use of a set of resonant controllers in parallel allows the rejection of disturbances and reference tracking at selected frequencies. However, a high number of resonant compensators makes control tuning more complicated. Furthermore, it is necessary to take into account the ranges of the load and its variation in the problem formulation. Although there is a large number of published results dealing with DER system control, topics that still deserve further investigation are the proposition of controller design methodologies according to specific standards and the use of conditions based on Lyapunov functions for a more rigorous analysis of the stability and robust performance of the closed-loop system.

In the literature, different controllers have been applied to control the distributed energy system based on the inverter with an LC filter. In Reference [5], PID (Proportional Integral Derivative) controllers were proposed. However, these ones were based on the voltage RMS value to obtain the feedback signal. Thus, their use solves the problem of voltage amplitude, but, in general, it does not solve the problem of distortion of the output voltage waveform [6]. A cascaded voltage/current control based on a virtual impedance concept has been developed in [7]. This topology simultaneously improves the injected grid current and local load voltage in island and grid connected microgrids. An optimized proportional resonant (PR) controller taking account of the computational delay from the digital control system was proposed in [8]. A controller based on a polytopic model was proposed in [9] for the current controller development by taking into account the grid inductance variations and soft saturation. To minimize output current ripples and reduce the steady-state current deviation, a quasi proportional resonant controller was proposed in [10]. Generally, the proportional resonant controller parameters are tuned manually. However, in [8], the authors proposed an optimization tuning of the RCs controller. An adaptive PR controller was proposed in [11] to ensure a current tracking under grid parameter variations. The authors used a fourth order filter with the adaptive RC to estimate online the resonance frequency, and then the grid impedance variations. In order to reduce the steady state current error, a modified proportional resonant controller used for the inner current control loop and indirect vector controller at the rotating reference frame was proposed in [12]. A controller to compensate the resonance phenomena without knowing the system parameters and without affecting the controller bandwidth in island microgrids was proposed in [13]. A three degrees of freedom cascaded voltage/current controller based on $\mathcal{H}_\infty$ theory was proposed in [14,15]. The main advantage of this controller is to ensure the stability of the closed-loop system and the robustness against parameter uncertainties.

To solve the above problems, the proposed method in this paper is based on the solution of optimization problems addressed by linear matrix inequalities (LMIs) formulation, which allows to concatenate the parameters of different resonant compensators acting in different frequencies. The proposed design method includes an easy solution to synthetizing the gains of the resonant controller, for a given capacity of the applied load range to distributed generation system units in the microgrid. The determination of gains is formalized through a convex optimization problem under a set of linear matrix inequalities constraints. In this paper, state feedback is used in conjunction with multiple resonant compensators that ensure the elimination of the steady state error and guarantee the rejection of multiple harmonics of the fundamental frequency produced by non-linear loads. In order to ensure the desired performances, the LMIs constraints construction is performed using a performance criterion throughout a robust regional poles placement [16,17].

Following this paper, a mathematical modeling of a DER system is proposed in Section 2. In Section 3 the proposed controller is presented, as well as the augmented model of the DER system and proportional resonant controllers including harmonic compensator and disturbance. Followed by the fundamental concepts for stability by Lyapunov and D-stability based on LMIs, which are discussed in Section 4. The simulation results are presented in Section 5, where the obtained values of the controller gains are given as well as considered DER system parameters, the controller discretization method, and the structure of the digital implementation of the controller. In Section 5, the results obtained with the use of the controller designed for the simulated system are discussed, with the description of the tests used and discussion for each of the results. Finally, in Section 6, the conclusions reached from the developments and the simulation of this contribution are presented, as well as the definition of possible lines of development for the future.

2. Mathematical Modeling of DER System

2.1. System Description

In the distributed energy system, the conditioning of the output voltage is performed by a DC-AC voltage inverter together with a second-order LC Low Pass filter, as shown in Figure 1. This figure shows the electrical diagram of the inverter used and the polarity of the currents involved. The model to be presented to the distributed energy system disregards the influence of the use of batteries and filter capacitors. The switching circuit (or switching) composed of the IGBT (Insulated Gate Bipolar Transistor) switches fed by the DC link voltage and the PWM (Pulse Width Modulation) modulator is modeled by the average value of the voltage applied to the LC low-pass filter and which will be represented by the input voltage $V_{DER} = k_{PWM}\ V_{CC}$, where k_{PWM} is the gain of the modulator and V_{CC} the voltage of the DC link.

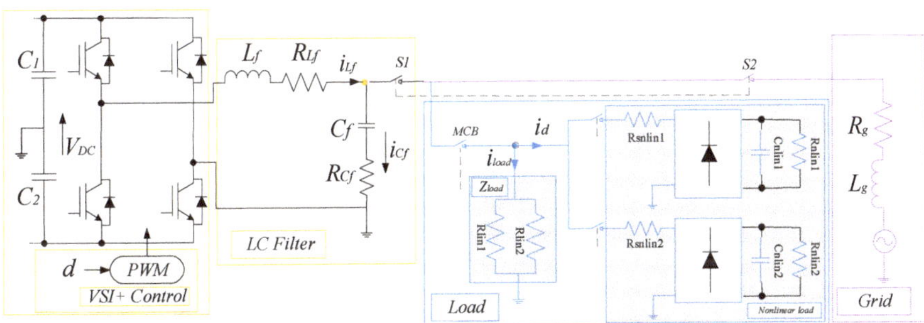

Figure 1. Distributed energy system circuit.

2.2. Linear Load Model Considerations

From Figure 1, the linear load model considers two fundamental factors that serve to understand the equations of modeling the DER system with a connected load.

1. Considers the Y_{load} admittance as a time-varying parameter, where only the admittance values are known considering the two generic DER operating regimes: empty operation (minimum admittance) and nominal (maximum admittance).
2. Considers current disturbances such as: sags, swell, notch or peak currents among other disturbances, and are modeled by the disturbance current i_d [18]. The i_d current is also an input to the DER system, but it is not controllable and its value will depend on the load, unlike the V_{DER} signal. In this way, the i_d current is considered as a disturbance to the system and its presence is seen as a source of loss of system performance.

The load admittance value is in the range of $[0.0001; 0.2\,S]$, that is:

$$Y_{\text{load}} = \frac{1}{Z_{\text{load}}}, Y_{\text{load}} \in \Delta := \{Y_{\text{load}} : 0.0001\,S \leq Y_{\text{load}} \leq 0.2\,S\} \tag{1}$$

2.3. State Space Modelling of DER System

The representation of the system in the state space describes the dynamic behavior of all the energy storage elements in the distributed energy system, which in this case are the inductor L_f and capacitor C_f belonging to the inverter filter. To better understand the physical relationships between the components of the inverter, Figure 2 shows the block diagram representation of the inverter that shows the relationship between the capacitor voltage and the inductor current. The block diagram facilitates understanding and later description by state variables. As the load connected to the DER system can be linear or non-linear, this paper considers a representation of the load current in terms of two distinct components. The first current, derived from the admittance Y_{load}, which represents a linear load that consumes a current i_{load}. The second added to i_{load} is an i_d current, which in this paper represents disturbances or disturbances to the value of the load applied to the distributed energy system. This current represents in a simplified way the model of non-linear loads connected to a DER system [19,20].

The initial methodology to compose the state matrices requires the determination of the differential equations that govern the behavior of the voltages and currents involved in the filter, according to the polarity defined in Figure 1. Basically, two equations are used, one referring to the mesh equation and the other one to the node equation as shown in (2).

$$\begin{aligned} V_{DER} &= v_{\text{load}} + v_{Lf} + i_{Lf}R_{Lf} \\ i_{Lf} &= i_{Cf} + i_{\text{load}} + i_d \end{aligned} \tag{2}$$

Knowing that the relationship between voltage and current in the inductor and capacitor are given by:

$$\begin{aligned} i_{Cf} &= C_f \frac{dv_{Cf}}{dt} \\ v_{Lf} &= L_f \frac{di_{Lf}}{dt} \end{aligned} \tag{3}$$

The voltage drop in the internal resistance of the inductor is proportional to the current flowing through it, as well as the voltage drop in the internal resistance of the capacitor. In this way, Equation (4) is defined.

$$\begin{aligned} v_{R_{Lf}} &= R_{Lf} i_{Lf} \\ v_{R_{Cf}} &= R_{Cf} i_{Cf} \end{aligned} \tag{4}$$

and taking into account the voltage drop in the resistance referring to the losses in the capacitor, we obtain:

$$\begin{aligned} v_{\text{load}} &= v_{Cf} + v_{R_{cf}} \\ v_{\text{load}} &= v_{Cf} + i_{Cf} R_{Cf} \end{aligned} \tag{5}$$

From (3), the current is obtained at the series impedance of the capacitor, reaching the equation for voltage at the load:

$$v_{load} = v_{Cf} + C_f \frac{dv_{Cf}}{dt} R_{Cf} \quad (6)$$

By replacing the two equations defined in (3), we obtain the relationship defined in (7).

$$i_{Lf} = C_f \frac{dv_{Cf}}{dt} + v_{load} \cdot Y_{load} + i_d \quad (7)$$

And replacing the variable V_{load} of Equation (7) in Equation (6) we obtain Equation (8)

$$i_{Lf} = C_f \frac{dv_{Cf}}{dt} + Y_{\text{load}} \left(v_{Cf} + C_f \frac{dv_{Cf}}{dt} R_{Cf} \right) + i_d \quad (8)$$

We then arrive at an equation where there are only the measured variables related to each other as presented in (9).

$$\frac{dv_{Cf}}{dt} \left(C_f + Y_{\text{load}} \cdot C_f R_{Cf} \right) = i_{Lf} - Y_{\text{load}} \cdot v_{Cf} - i_d \quad (9)$$

As there is the presence of the capacitor voltage and its derivative, the following manipulation can be done to arrive at the first state Equation (10), related to the capacitor voltage.

$$\frac{dv_{Cf}}{dt} = \frac{\frac{1}{Y_{\text{load}}}}{C_f \left(\frac{1}{Y_{\text{load}}} + R_{Cf} \right)} i_{Lf} - \frac{1}{C_f \left(\frac{1}{Y_{\text{load}}} + R_{Cf} \right)} v_{Cf} - \frac{\frac{1}{Y_{\text{load}}}}{C_f \left(\frac{1}{Y_{\text{load}}} + R_{Cf} \right)} i_d \quad (10)$$

Having obtained the first equation of state from the law of nodes applied to the system, the process for obtaining the second equation of state will now be started based on the law of meshes, given by Equation (11).

$$V_{DER} = v_{Lf} + v_{R_{Lf}} + v_{Cf} + v_{R_{Cf}} \quad (11)$$

Using the study already carried out, the manipulation of the equation begins by substituting (3), and (4), leading expression (12).

$$V_{DER} = L_f \frac{di_{Lf}}{dt} + R_{Lf} i_{Lf} + v_{Cf} + R_{Cf} i_{Cf} \quad (12)$$

Replacing (12) in (2), it takes the following expression:

$$V_{DER} = L_f \frac{di_{Lf}}{dt} + R_{Lf} i_{Lf} + v_{Cf} + C_f R_{Cf} \frac{dv_{Cf}}{dt} \quad (13)$$

From the replacement of Equation (10) in Equation (13) we have the expression in (14).

$$V_{DER} = L_f \frac{di_{Lf}}{dt} + R_{Lf} i_{Lf} + v_{Cf} + C_f R_{Cf} \left[\frac{\frac{1}{Y_{\text{load}}} i_{Lf}}{C_f \left(\frac{1}{Y_{\text{load}}} + R_{Cf} \right)} - \frac{v_{Cf}}{C_f \left(\frac{1}{Y_{\text{load}}} + R_{Cf} \right)} - \frac{\frac{1}{Y_{\text{load}}} i_d}{C_f \left(\frac{1}{Y_{\text{load}}} + R_{Cf} \right)} \right] \quad (14)$$

Factoring and manipulating algebraically (14) we obtain (15).

$$V_{DER} = L_f \frac{di_{Lf}}{dt} + i_{L_f} \left[R_{Lf} + \frac{\frac{1}{Y_{\text{load}}} R_{Cf}}{\left(\frac{1}{Y_{\text{load}}} + R_{Cf} \right)} \right] + v_{Cf} \left[1 - \frac{R_{Cf}}{\left(\frac{1}{Y_{\text{load}}} + R_{Cf} \right)} \right] - i_d \frac{\frac{1}{Y_{\text{load}}} R_{Cf}}{\left(\frac{1}{Y_{\text{load}}} + R_{Cf} \right)} \quad (15)$$

Finally, Equation (16) comes from the state of the inductor current, which is extracted through the algebraic manipulation made in Equation (15).

$$\frac{di_{Lf}}{dt} = \frac{V_{DER}}{L_f} - i_{Lf}\frac{1}{L_f}\left[R_{Lf} + \frac{\frac{R_{Cf}}{Y_{\text{load}}}}{\left(\frac{1}{Y_{\text{load}}}+R_{Cf}\right)}\right] - v_{Cf}\frac{1}{L_f}\left[1 - \frac{R_{Cf}}{\left(\frac{1}{Y_{\text{load}}}+R_{Cf}\right)}\right] + i_d\frac{\frac{1}{Y_{\text{load}}}R_{Cf}}{\left(\frac{1}{Y_{\text{load}}}+R_{Cf}\right)} \tag{16}$$

Defining the current in the inductor as state and the voltage in the capacitor as another state, we obtain the following representation by the state variables:

$$\begin{aligned} \dot{x}(t) &= Ax(t) + Bu(t) + B_w w(t) \\ z(t) &= Cx(t) \end{aligned} \tag{17}$$

where $x(t) = \begin{bmatrix} x_1 & x_2 \end{bmatrix}^T = \begin{bmatrix} i_{Lf} & v_{Cf} \end{bmatrix}^T$ is the state vector, $w(t) = i_d$ is the external disturbance signal, $u(t) = V_{DER}$ is the control signal (PWM output), $z(t) = v_{Cf}$ is the output of the system, $C = \begin{bmatrix} 0 & 1 \end{bmatrix}$, and it is observed in the model, that the current i_d is not a variable of known characteristic and is modeled as a disturbance external to the model.

$$A = \begin{bmatrix} -\left(\frac{R_{Lf}}{L_f} + \frac{\frac{1}{Y_{\text{load}}}R_{Cf}}{L_f\left(\frac{1}{Y_{\text{load}}}+R_{Cf}\right)}\right) & \frac{-\frac{1}{Y_{\text{load}}}}{L_f\left(\frac{1}{Y_{\text{load}}}+R_{Cf}\right)} \\ \frac{\frac{1}{Y_{load}}}{C_f\left(\frac{1}{Y_{\text{load}}}+R_{Cf}\right)} & \frac{-1}{C_f\left(\frac{1}{Y_{\text{load}}}+R_{Cf}\right)} \end{bmatrix} B = \begin{bmatrix} \frac{1}{Lf} \\ 0 \end{bmatrix}, B_w = \begin{bmatrix} \frac{\frac{1}{Y_{\text{load}}}R_{Cf}}{L_f\left(\frac{1}{Y_{\text{load}}}+R_{Cf}\right)} \\ \frac{-\frac{1}{Y_{load}}}{C_f\left(\frac{1}{Y_{load}}+R_{Cf}\right)} \end{bmatrix} \tag{18}$$

The DER system considered in this paper uses capacitors with low resistance RCf. Thus, in order to simplify the model obtained in (17) and (18), it is considered that $R_{Cf} = 0$. Thus, starting from (18), the following state matrices are obtained for the system defined in (17):

$$\begin{aligned} A\left(Y_{\text{load}}\right) &= \begin{bmatrix} -\left(\frac{R_{Lf}}{Lf}\right) & \frac{-1}{Lf} \\ \frac{1}{Cf} & \frac{-Y_{\text{load}}}{Cf} \end{bmatrix}, B = \begin{bmatrix} \frac{1}{Lf} \\ 0 \end{bmatrix} \\ B_w &= \begin{bmatrix} 0 \\ \frac{-1}{C_f} \end{bmatrix}, \quad C = \begin{bmatrix} 0 & 1 \end{bmatrix} \end{aligned} \tag{19}$$

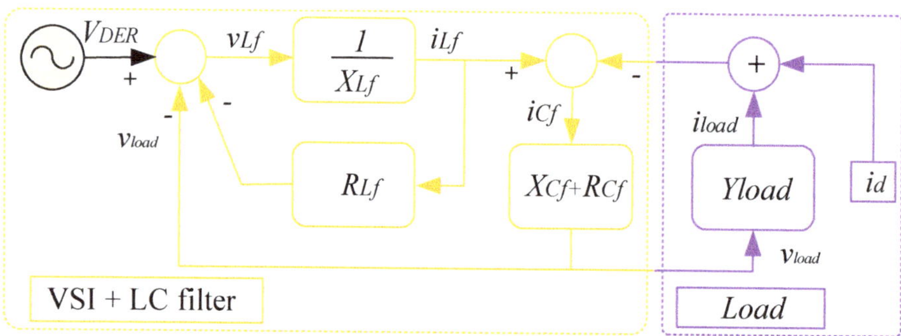

Figure 2. Inverter block diagram with filter and load.

3. Proportional Resonant Controller Model

In order to develop an effective control method to ensure the tracking of a sinusoidal reference, as well as the rejection of disturbances related to non-linear loads and harmonic compensation, new methodologies have been developed. In this work, proportional resonant controllers [21], based on the Internal Model Principle [22], are used. The proportional resonant compensators impose poles on the imaginary axis at frequencies equal to those of the reference, and show infinite gain in the resonance frequency. As real loads do not drain current linearly (with a waveform equal to the waveform of the applied voltage), the use of only one compensator at the fundamental frequency is not sufficient to keep a satisfactorily low THD. In particular, for DERs, typical non-linear loads can be seen as current disturbances in the frequency of the output voltage and in its multiples. The rejection of this type of disturbance will be fundamental for the good performance of a DER system. Consequently, a set of resonant compensators is used to reject each multiple harmonic frequency of the fundamental, by allocation of gain in multiple frequencies [23,24]. The following transfer function is usually used in the implementation of the ideal resonant controller.

$$C_{Rh}(s) = \frac{s^2}{s^2 + \omega_h^2} \tag{20}$$

where ω_h is the frequency of the signal to be followed or rejected. In some applications a damping factor ξ is inserted at the poles of Equation (20) to avoid discrete implementation problems regarding the position of resonant poles at the edge of the unit circle [25]. Given a resonant controller with transfer function:

$$C_{Rh}(s) = \frac{s^2}{s^2 + 2\xi\omega_h s + \omega_h^2} \tag{21}$$

Note that by making $\xi = 0$ the controller (21) reduces to (20). Controllers like this allocate imaginary pole pairs at the frequency of the reference to be traced and the frequencies of the disturbances to be rejected. To represent the multiple compensators, Equation (21) that models the resonant compensator in general is repeated in parallel in the block diagram of the control system, as shown in Figure 3. The multiple-resonant gain controller in (21) with proportional gain can be rewritten as

$$G_{PR_h}(s) = k_2 + \sum_{h=1}^{n} \frac{k_{h+1} + k_{h+2}s}{s^2 + 2\xi_h\omega_h s + \omega_h^2} \tag{22}$$

where k_{h+1} and k_{h+2} are the gains to be determined for each mode and k_2 is a direct transmission term applied to the input signal from the controller.

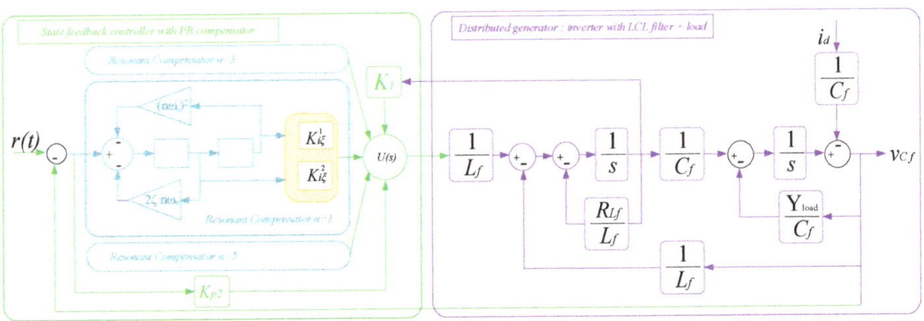

Figure 3. Block diagram of closed loop system for proportional resonant controller and inverter with LCL filter.

Controller Based on the Principle of the Internal

Model The design of controllers that guarantee follow-up of references and the rejection of disturbances is of great practical interest, being one of the main study topics for a number of authors over the years. The Principle of the Internal Model (IMP) [22] is an essential theoretical result for the design of a control system aiming at the follow-up of reference signals and rejection of disturbance signals. Its fundamental idea is to generate, within the control loop, a signal with the same characteristics as the signals to be followed and/or rejected. The closed-loop system must contain all persistent modes (which do not tend to zero at steady state) of the reference and disturbance signals in order to guarantee perfect tracking/rejection [26] . Using the principle of superposition, a controller based on the internal model principle simultaneously includes the reference and disturbance poles for the asymptotically stable system and without the cancellation of mentioned poles and zeros, the reference follow-up and rejection of disturbances with multiple harmonics, which occur in DERs with typical non-linear loads, with zero error in steady state. According to the principle of the internal model [22], in a feedback control system like the one in Figure 3, there is zero error in a steady state if the closed loop system is asymptotically stable and the poles of the system include the poles of reference to be tracked. In order to present the way in which the proportional resonant control tuning method was developed, initially it is necessary to obtain the state-space representation of the transfer function presented in Equation (22) [25,27], as in system (23):

$$\begin{aligned} \dot{\eta} &= A_{CR}\eta(t) + B_{CR}e(t) \\ y_{CR} &= C_{CR}\eta(t) + D_{CR}e(t) \end{aligned} \tag{23}$$

The representation of (23) in the state space is given by

$$\begin{cases} \dot{\eta}_r(t) = \begin{bmatrix} A_{CR_1} & \cdots & 0_2 \\ \vdots & \ddots & \vdots \\ 0_2 & \cdots & A_{CR_n} \end{bmatrix} \eta_r(t) + \begin{bmatrix} B_{CR_1} \\ \vdots \\ B_{CR_n} \end{bmatrix} e(t) \\ y_{CR}(t) = \begin{bmatrix} C_{CR_1} \ldots C_{CR_n} \end{bmatrix} \eta_r(t) + \begin{bmatrix} D_{CR} \end{bmatrix} e(t) \end{cases} \tag{24}$$

where $\eta_{CR}(t) \in \mathbb{R}^{2n}$ is the state vector of the multi-resonant controller, $e(t)$ is the input signal, $y_{CR}(t)$ is the output signal, and A_{CR}, B_{CR} and C_{CR} are matrices of appropriate size.

$$\begin{aligned} A_{CR_h} &= \begin{bmatrix} 0 & \omega_h \\ -\omega_h & -2\xi_h\omega_h \end{bmatrix}, \quad B_{CR_h} = \begin{bmatrix} 0 \\ 1 \end{bmatrix} \\ C_{CR_h} &= \begin{bmatrix} k_{h+1} & k_{h+2} \end{bmatrix}, \quad D_{CR_h} = [k_2] \end{aligned} \tag{25}$$

for each resonant mode, that is, evaluated for each pair (ξ_h, ω_h), $h = 1 \ldots n$. The controller based on the internal model principle has internal variables as output variables, which facilitates the design of the controller gains by full state feedback, since all state variables are available for control, as shown in Figure 3. The augmented system is composed only of the plant and proportional compensators based on internal model principle in the frequency of the fundamental, ω_0, and the characteristic of rejection of load disturbances in the other harmonic frequencies. Generalizing the representation of the plant with the controller, it can be written that :

$$\begin{cases} \dot{\tilde{x}}(t) = \tilde{A}\tilde{x}(t) + \tilde{B}_u u(t) + \tilde{B}_w w(t) + \tilde{B}_r r(t) \\ \tilde{y}(t) = \tilde{C}\tilde{x}(t) \end{cases} \tag{26}$$

where $w(t) = i_d$

$$\tilde{x}(t)=\begin{bmatrix} x(t)\\ \eta_1(t)\\ \eta_2(t)\\ \vdots\\ \eta_n(t)\end{bmatrix}\quad \tilde{A}=\begin{bmatrix} A(Y_{load}) & 0_{2\times2} & 0_{2\times2} & \cdots & 0_{2\times2}\\ -B_{CR1}C & A_{CR1} & 0_{2\times2} & \cdots & 0_{2\times2}\\ -B_{CR2}C & 0 & A_{CR2} & \cdots & 0_{2\times2}\\ \vdots & \vdots & \vdots & \ddots & \vdots\\ -B_{CRn}C & 0_{2\times2} & 0_{2\times2} & \cdots & A_{CRn}\end{bmatrix}, \tilde{B}_u=\begin{bmatrix} B\\ 0_{2\times1}\\ 0_{2\times1}\\ \vdots\\ 0_{2\times1}\end{bmatrix}$$
$$\tilde{C}=\begin{bmatrix} C & 0_{1\times2} & \cdots & 0_{1\times2}\end{bmatrix}, \tilde{B}_w=\begin{bmatrix} B_w\\ 0_{2\times1}\\ 0_{2\times1}\\ \vdots\\ 0_{2\times1}\end{bmatrix},\quad \tilde{B}_r=\begin{bmatrix} 0_{2\times1}\\ B_{CR1}\\ B_{CR2}\\ \vdots\\ B_{CRh}\end{bmatrix} \tag{27}$$

4. Robust Controller Design Based on LMIs

$\mathcal{D}$-Stability

The chosen $\mathcal{D}$-stability region is given by the intersection of a semi-plane sector (ρ) to guarantee a lower limit on settling time. Disc sector (σ) is used as an upper limit on settling time. It restricts the magnitude of the eigenvalues indirectly and imposes a limitation on the control effort. LMI regions [28] are defined as regions of the complex plane that can be described in terms of characteristic functions as follows:

$$\mathcal{D}=\{s\in\mathbb{C}: f_{\mathcal{D}}(s)<0\}, f_{\mathcal{D}_{\sigma,\rho,\theta}}(s)=L+sM+\bar{s}M^{\mathrm{T}} \tag{28}$$

Lemma 1 below presents a condition that ensures the quadratic $\mathcal{D}$-stability of the closed loop system (26) in a given LMI region.

Lemma 1. *Quadratic stability of $\mathcal{D}$-Region [28]. Consider the LMI region with a characteristic function (28). Additionally, consider that $M=M_1^TM_2$, where both M_1 and M_2 are matrices with k lines and rank per complete line. The system (26) presents quadratic $\mathcal{D}$-stability with region (28) if there are real symmetric matrices $\mathcal{P}>0$ and $\mathcal{U}>0$ such that:*

$$\begin{bmatrix} L\otimes\mathcal{P}+He\{M\otimes(\mathcal{P}(\tilde{A}+\tilde{B}_uK))\} & M_1^T\otimes(\mathcal{P}B_\Delta) & (M_2^T\mathcal{U})\otimes C_\Delta^T\\ M_1\otimes(B_\Delta^T\mathcal{P}) & -\mathcal{U}\otimes I & 0\\ (\mathcal{U}M_2)\otimes C_\Delta & 0 & -\mathcal{U}\otimes I\end{bmatrix}<0 \tag{29}$$

When the LMI region of interest is formed by the intersection of other LMI regions, quadratic $\mathcal{D}$-stability is guaranteed when the same matrix P meets (26) simultaneously with the characteristic functions of each region [29].

In particular, the region of our work can be described as $\mathcal{D}=\mathcal{D}_\sigma\cap\mathcal{D}_\rho$, where

$$\begin{aligned}&\mathcal{D}_\sigma=\{\delta_i\in\mathbb{C}:\Re(\delta_i)<-\sigma,\sigma\geqslant0,\sigma\in\mathbb{R}\}, L_\sigma=2\sigma; M_\sigma=1\\ &\mathcal{D}_\rho=\{\delta_i\in\mathbb{C}:|\delta_i+q|<\rho,\rho>0,\rho\in\mathbb{R}\}, L_\rho=\begin{bmatrix}-\rho & q\\ q & -\rho\end{bmatrix}, M_\rho=\begin{bmatrix}0 & 1\\ 0 & 0\end{bmatrix}\end{aligned} \tag{30}$$

The following theorem details the conditions in the form of LMIs for the positioning of the eigenvalues of $\tilde{A}+B_\Delta\Delta(t)C_\Delta+\tilde{B}_uK$ in $\mathcal{D}=\mathcal{D}_\sigma\cap\mathcal{D}_\rho$

Remark. *One can notice that the quadratic $\mathcal{D}$-stability is wider than the $\mathcal{H}_\infty$ condition. Indeed, the quadratic $\mathcal{D}$-stability allows to guarantee the stability of the uncertain system as well as the $\mathcal{H}_\infty$ performance. We refer the interested reader to [28] and the references therein.*

Theorem 1. *When real scalars σ and ρ are known, the closed-loop system $\dot{\tilde{x}}(t) = \left(\tilde{A}\left(Y_{load}(t)\right) + \tilde{B}_u K\right)\tilde{x}(t)$ is asymptotically stable for all $\left(Y_{load}\right)_{\min} \leq Y_{load}(t) \leq \left(Y_{load}\right)_{\max}$, and the poles are located in the region $S(\sigma,\rho)$, if there exists a symmetric positive definite matrices $Q \in R^{(n+2h)\times(n+2h)}$, $W \in R^{1\times(n+2h)}$ such that the following LMI are feasible:*

$$\begin{aligned} &\tilde{A}\left(Y_{load}\right)_{\min} Q + Q\tilde{A}^T\left(Y_{load}\right)_{\min} + \tilde{B}_u W + \tilde{B}_u^T W^T + 2\sigma Q < 0 \\ &\tilde{A}\left(Y_{load}\right)_{\max} Q + Q\tilde{A}^T\left(Y_{load}\right)_{\max} + \tilde{B}_u W + \tilde{B}_u^T W^T + 2\sigma Q < 0 \end{aligned} \quad (31)$$

$$\begin{aligned} &\begin{bmatrix} -\rho Q & \tilde{A}\left(Y_{load}\right)_{\min} Q + \tilde{B}_u W \\ Q\tilde{A}^T\left(Y_{load}\right)_{\min} + W^T\tilde{B}_u^T & -\rho Q \end{bmatrix} < 0 \\ &\begin{bmatrix} -\rho Q & \tilde{A}\left(Y_{load}\right)_{\max} Q + \tilde{B}_u W \\ Q\tilde{A}^T\left(Y_{load}\right)_{\max} + W^T\tilde{B}_u^T & -\rho Q \end{bmatrix} < 0 \end{aligned} \quad (32)$$

With: He $\left\{\tilde{A}Q + \tilde{B}_u W\right\}$ and He $\left\{\tilde{A}Q - \tilde{B}_u W\right\}$ represent the hermitian block $\left(\tilde{A}\left(Y_{load}\right)_{\min}^{\max} Q + Q\tilde{A}^T\left(Y_{load}\right)_{\min}^{\max} + \tilde{B}_u W + \tilde{B}_u^T W^T\right)$, $\left(\tilde{A}\left(Y_{load}\right)_{\min}^{\max} Q - Q\tilde{A}^T\left(Y_{load}\right)_{\min}^{\max} + \tilde{B}_u W - \tilde{B}_u^T W^T\right)$, respectively.

$$\begin{aligned} &\min_{Q,W} \gamma \quad \textit{subject to:} \\ &Q = Q^T > 0, \textit{ optimization problems in } (29) \textit{ and } (30) \end{aligned} \quad (33)$$

5. Simulation Results and Discussion

Using the robust control law presented in the previous sections for synthesizing gains of the distributed energy system, this section groups results from simulations tests. The numerical values of electrical components, LCL filter and resistances and capacitances parameters associated with each load situation for design purposes of the DER controller are listed in Table 1. The reference signal to be followed by the output voltage is sinusoidal with an RMS value of 127 V and frequency of 60 Hz. The inverter switching frequency is 21.6 kHz, 360 times greater than the frequency of the reference signal. The simulation environment is given by Matlab/Simulink for the DER system with linear and non-linear test loads, including the generation of PWM signal for inverter switching in a half bridge and DER control loop, with a controller properly discretized by zero-order hold with a sampling frequency of 21.6 kHz as shown in Figure 4. The LMI parameters chosen in this paper to improve the dynamic response and stability in steady-state are shown in Table 2.

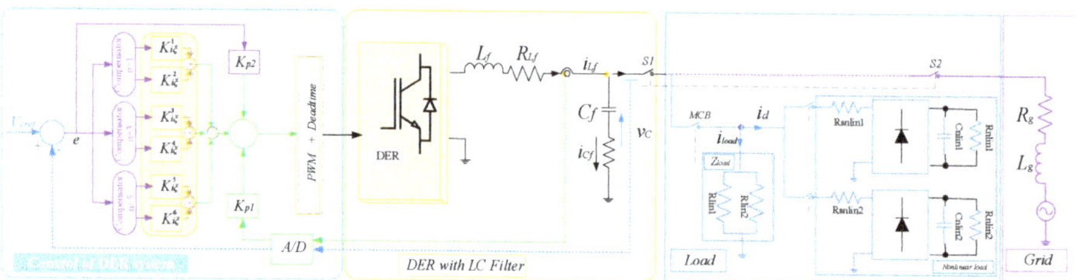

Figure 4. General scheme of the distributed energy resource (DER) system with the proposed control.

Table 1. DER system parameters.

Parameter	Value
Nominal power	5.2 KVA
RMS output voltage	127 V
Frequency	60 Hz
Output power factor	0.7
Inductance	$L_f = 1$ mH
Parasitic resistance of the inductor	$R_{Lf} = 0.015\,\Omega$
Capacitance	$C_f = 250\,\mu$F
Capacitive parasitic resistance (considered)	$R_{Cf} = 0\,\Omega$
Uncertain linear load impedance	$Y_{\min}, Y_{\max} = [5; 10,000]\,\Omega$
DC bus capacitances	$C_1, C_2 = 6600\,\mu$F
DC bus voltage	$V_{cc} = 520$ V
Load parameters	$R_{lin1} = 32.92\,\Omega$, $R_{lin2} = 8.23\,\Omega$ $R_{snlin1} = 0{,}73\,\Omega$, $R_{snlin2} = 0.75\,\Omega$ $R_{nlin1} = 37.2\,\Omega$, $R_{nlin2} = 16.5\,\Omega$ $C_{nlin1} = 3010\,\mu$F, $C_{nlin2} = 9010\,\mu$F

Table 2. Linear matrix inequalities (LMI) parameters used for controller design.

Parameter	Value
σ (Sigma position)	100
ρ (radius of the circle)	20,000
c (center of circle)	0

Figures 5 and 6 show the voltage and current curve under load steps in the period of two seconds considering additional load steps of 20% at $t = 0.2$ s and of 80% at $t = 0.6$ s. Following by a subtractive step of $t = 80\%$ at $t = 1.4$ s and, finally, a subtractive step of 20% at $t = 1.8$ s. In Figure 5a, a cycle (16.67 ms) of the voltage (127 V RMS) and output current (Figure 6a) of the DER system with 100% linear load. In this figure it is evident how the voltage curve fits almost perfectly over the reference curve. Similarly for the current curve, which has an almost sinusoidal appearance with and without controller with fundamental frequency and a harmonics compensator (HC). Figures 5b and 6b show voltage and output current curves from the DER system with 100% non-linear load with HC, differently as for linear load, in non-linear load, the deformation of the current curve is more pronounced, which is caused by this type of load. Figures 5c and 6c show the simulation results for voltage and current without HC in the controller; from these figures we can seen that the non-linear load creates a voltage deviation voltage that increases proportionally to the connected step load. In the linear and non-linear load with HC, the shape of the current is slightly deformed compared with 100% non-linear load without HC. However, both non-linear loads present small transients and overshoots for additive and subtractive load steps contrary to the linear load, which has voltage and current responses that are perfectly sinusoidal.

Figure 7 shows the values of the empty RMS voltage, with 20% and 100% linear load, and with 20% and 100% non-linear load, respectively, of the DER system. In this study, we take account of the steady state voltage without the transients of connected and disconnected loads. For the linear load, we observe that the RMS value of the voltage remains satisfying the most stringent standard, which is also the case for the non-linear load with the controller use HC. However, for the non-linear load without HC, when the addition of 100% non-linear load occurs, the RMS value of the voltage slightly exceeds (2%) the limit of the IEEE 944 standard. Figure 8 shows the THD of the DER system output voltage for all additive and subtractive steps with different periods for linear and non-linear loads. In the cases where the PR controllers are used with harmonic compensators, the THD with 25% and 100% linear/non-linear loads varying meet the IEEE and IEC standards for the proposed LMI approaches, as shown in Figure 8a,b. However, when the PR controllers are used without harmonic compensators, the THD with 25% non-linear load satisfied the IEC 62040-3 standard and IEEE 944 standard norms but did not satisfy the norms with 100% non-linear load. For each case, the THD values of the DER system with empty, with 20% and 100% linear/non-linear load are shown in Table 3.

It is observed in Figure 9a–c that the system is running empty and, when the step non-linear load increases (with and without HC in controller), the voltage deviation increases, which confirms the results shown in Figure 5. The increasing of voltage deviation is more important when not using a harmonics compensator in the controller. However, when the system undergoes a subtraction of the non-linear load step, the voltage deviation decreases. It is important to note that this graph does not take into account load insertion or removal transients. Our interest in this case in assessing this graph is the voltage in steady state. However, it is possible to observe that transients appear in the voltage deviation value due to the loads connection/disconnection. These transients do not impact the proposed control system performance because of the output voltages' small variation in the amplitude and the fast recovery time. In fact, transients remain in compliance with IEEE 944 standard limits.

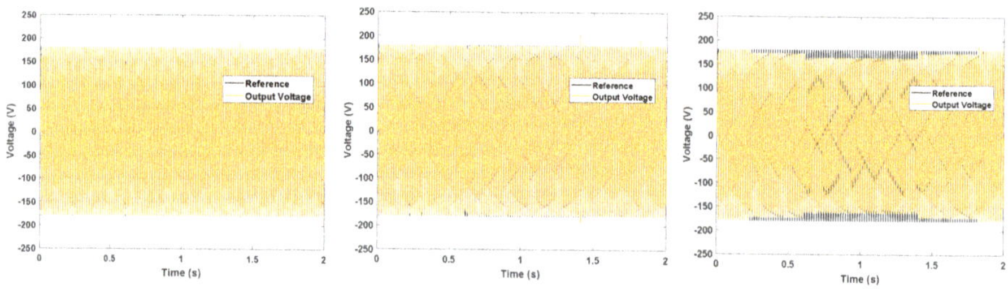

(a) (b) (c)

Figure 5. DER output voltage curve different loads. (**a**) 100% linear load with HC. (**b**) 100% non-linear load with HC. (**c**) 100% non-linear load without harmonics compensator (HC).

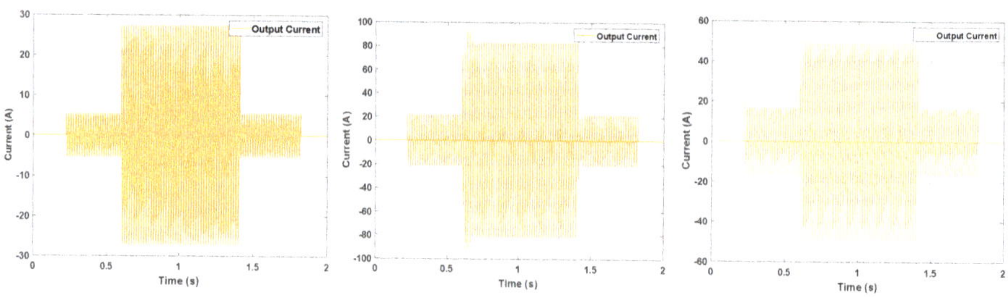

(a) (b) (c)

Figure 6. DER system output current curve. (**a**) 100% linear load with HC. (**b**) 100% non-linear load with HC. (**c**) 100% non-linear load without HC.

Table 3. THD voltage with different load steps.

Operation Mode	Empty	Load 20%	Load 100%
Linear load	0.09%	0.088%	0.087%
Non-linear load with HC	0.062%	0.8107%	2.328%
Non-linear load without HC	0.12%	3.547%	9.71%

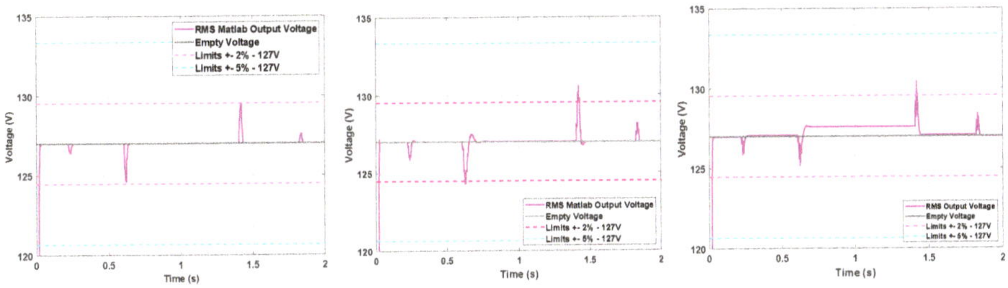

(a) (b) (c)

Figure 7. RMS value of DER system output voltage. (**a**) 100% linear load with HC. (**b**) 100% non-linear load with HC. (**c**) 100% non-linear load without HC.

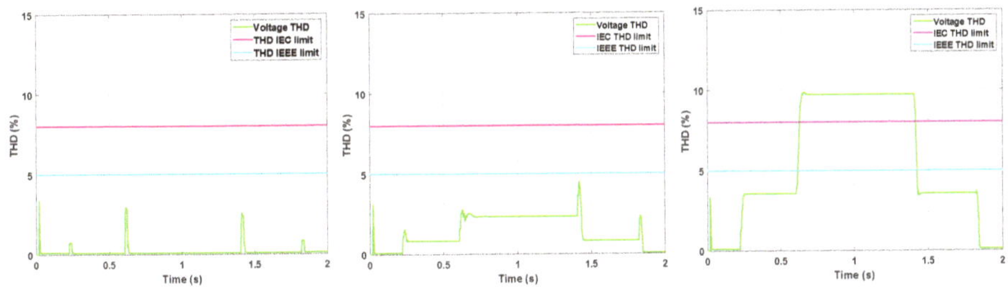

(a) (b) (c)

Figure 8. THD value of DER system output voltage. (**a**) 100% linear load with HC. (**b**) 100% non-linear load with HC. (**c**) 100% non-linear load without HC.

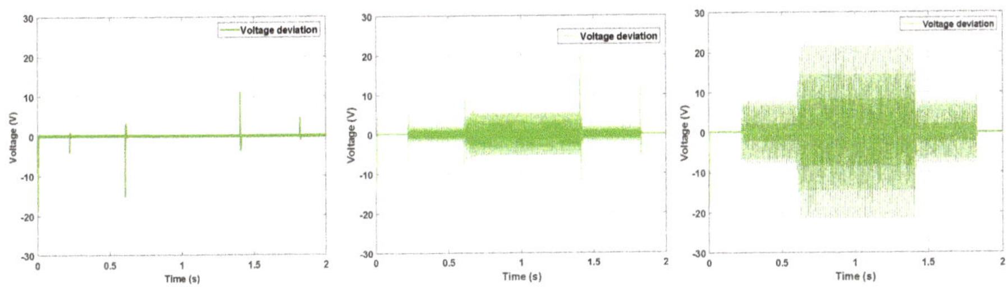

(a) (b) (c)

Figure 9. DER system output voltage deviation. (**a**) 100% linear load with HC. (**b**) 100% non-linear load with HC. (**c**) 100% non-linear load without HC.

6. Conclusions

The distributed generation resource systems are being used in the micro-grid applications and, therefore, the concern with their reliability grows as the currents requested from DER systems have a non-linear character. In this paper, a multiple resonant compensator strategy through the robust control theory is proposed for guaranteeing stability and performance for different types of loads including the problem of harmonic cancellation and attenuation of disturbances in the output voltage of DER systems caused by non-linear loads. The stability and performance criteria based on Lyapunov's theory are used and described using the formulation by linear matrix inequalities, where it was possible to establish design criteria for determining the gains of the multiple resonant compensators by allocating poles in order to guarantee a good dynamic performance and harmonic rejection. The simulation results prove that the controller studied in this work presents advantages in relation to the dynamic performance and in a steady state against load variations. In future work, an adaptive frequency system will be added to resonant compensators to increase robustness when a variation in the frequency of the reference signal is allowed. Another important point is to use Anti Windup compensators in parallel with the resonant to avoid saturation of the control signal.

Author Contributions: Author Contributions: Conceptualization, A.E.M.B., M.Z., S.B.E. and M.B.; Methodology, A.E.M.B., M.Z., S.B.E. and M.B.; Software, A.E.M.B.; Validation, A.E.M.B., M.Z., S.B.E. and M.B.; Formal Analysis, A.E.M.B., M.Z., S.B.E., K.B. and M.B.; Investigation, A.E.M.B.; Writing—Original Draft Preparation, A.E.M.B.; Writing—Review & Editing, A.E.M.B., M.Z., S.B.E., K.B. and M.B. All authors have read and agreed to the published version of the manuscript.

Funding: This research received no external funding.

Conflicts of Interest: The authors declare no conflict of interest.

References

1. Zia, M.F.; Elbouchikhi, E.; Benbouzid, M.; Guerrero, J.M. Microgrid Transactive Energy Systems: A Perspective on Design, Technologies, and Energy Markets. In Proceedings of the IECON 2019—45th Annual Conference of the IEEE Industrial Electronics Society, Lisbon, Portugal, 14–17 October 2019; Volume 1, pp. 5795–5800.
2. Zia, M.F.; Benbouzid, M.; Elbouchikhi, E.; Muyeen, S.M.; Techato, K.; Guerrero, J.M. Microgrid Transactive Energy: Review, Architectures, Distributed Ledger Technologies, and Market Analysis. *IEEE Access* **2020**, *8*, 19410–19432. [CrossRef]
3. Bouzid, A.M.; Chériti, A.; Sicard, P. H-infinity loopshaping controller design of micro-source inverters to improve the power quality. In Proceedings of the 2014 IEEE 23rd International Symposium on Industrial Electronics (ISIE), Istanbul, Turkey, 1–4 June 2014; pp. 2371–2378.
4. Eltawil, M.A.; Zhao, Z. Grid-connected photovoltaic power systems: Technical and potential problems—A review. *Renew. Sustain. Energy Rev.* **2010**, *14*, 112–129. [CrossRef]
5. Timbus, A.; Liserre, M.; Teodorescu, R.; Rodriguez, P.; Blaabjerg, F. Evaluation of current controllers for distributed power generation systems. *IEEE Trans. Power Electron.* **2009**, *24*, 654–664. [CrossRef]
6. Bouzid, A.M.; Golsorkhi, M.S.; Sicard, P.; Chériti, A. H∞ structured design of a cascaded voltage/current controller for electronically interfaced distributed energy resources. In Proceedings of the 2015 Tenth International Conference on Ecological Vehicles and Renewable Energies (EVER), Monte Carlo, Monaco, 31 March–2 April 2015; pp. 1–6.
7. Samavati, E.; Mohammadi, H. Simultaneous voltage and current harmonics compensation in islanded/grid-connected microgrids using virtual impedance concept. *Sustain. Energy Grids Netw.* **2019**, *20*, 100258. [CrossRef]
8. Husev, O.; Roncero-Clemente, C.; Makovenko, E.; Pimentel, S.P.; Vinnikov, D.; Martins, J. Optimization and Implementation of the Proportional-Resonant Controller for Grid-Connected Inverter With Significant Computation Delay. *IEEE Trans. Ind. Electron.* **2020**, *67*, 1201–1211. [CrossRef]
9. Osório, C.R.D.; Koch, G.G.; Pinheiro, H.; Oliveira, R.C.L.F.; Montagner, V.F. Robust Current Control of Grid-Tied Inverters Affected by LCL Filter Soft-Saturation. *IEEE Trans. Ind. Electron.* **2019**, *67*, 6550–6561. [CrossRef]
10. Ye, T.; Dai, N.; Lam, C.; Wong, M.; Guerrero, J.M. Analysis, Design, and Implementation of a Quasi-Proportional-Resonant Controller for a Multifunctional Capacitive-Coupling Grid-Connected Inverter. *IEEE Trans. Ind. Appl.* **2016**, *52*, 4269–4280. [CrossRef]
11. Khalfalla, H.; Ethni, S.; Al-Greer, M.; Pickert, V.; Armstrong, M.; Phan, V.T. An adaptive proportional resonant controller for single phase PV grid connected inverter based on band-pass filter technique. In Proceedings of the 2017 11th IEEE International Conference on Compatibility, Power Electronics and Power Engineering (CPE-POWERENG), Cadiz, Spain, 4–6 April 2017; pp. 436–441.
12. Kumar, N.; Saha, T.K.; Dey, J. Control, implementation, and analysis of a dual two-level photovoltaic inverter based on modified proportional-resonant controller. *IET Renew. Power Gener.* **2018**, *12*, 598–604. [CrossRef]
13. Saim, A.; Houari, A.; Guerrero, J.M.; Djerioui, A.; Machmoum, M.; Ahmed, M.A. Stability Analysis and Robust Damping of Multiresonances in Distributed-Generation-Based Islanded Microgrids. *IEEE Trans. Ind. Electron.* **2019**, *66*, 8958–8970. [CrossRef]

14. Bouzid, A.E.M.; Sicard, P.; Chaoui, H.; Cheriti, A. Robust three degrees of freedom based on $H\infty$ controller of voltage/current loops for DG unit in micro grids. *IET Power Electron.* **2019**, *12*, 1413–1424. [CrossRef]
15. Bouzid, A.M.; Sicard, P.; Chériti, A.; Bouhamida, M. H_∞ loopshaping controller design of micro-source inverters. In Proceedings of the 2014 IEEE 27th Canadian Conference on Electrical and Computer Engineering (CCECE), Toronto, ON, Canada, 4–7 May 2014; pp. 1–6.
16. Zerrougui, M.; Darouach, M.; Boutat-Baddas, L.; Souley-Ali, H. H_∞ filtering for singular bilinear systems with application to a single-link flexible-joint robot. *Int. J. Control Autom. Syst.* **2014**, *12*, 590–598. [CrossRef]
17. Zemouche, A.; Zerrougui, M.; Boulkroune, B.; Rajamani, R.; Zasadzinski, M. A new LMI observer-based controller design method for discrete-time LPV systems with uncertain parameters. In Proceedings of the 2016 American Control Conference (ACC), Boston, MA, USA, 6–8 July 2016; pp. 2802–2807.
18. Thirumala, K.; Pal, S.; Jain, T.; Umarikar, A.C. A classification method for multiple power quality disturbances using EWT based adaptive filtering and multiclass SVM. *Neurocomputing* **2019**, *334*, 265–274. [CrossRef]
19. Bouzid, A.M.; Sicard, P.; Paquin, J.; Yamane, A. A robust control strategy for parallel-connected distributed generation using real-time simulation. In Proceedings of the 2016 IEEE 7th International Symposium on Power Electronics for Distributed Generation Systems (PEDG), Vancouver, BC, Canada, 27–30 June 2016; pp. 1–8.
20. Bouzid, A.E.M.; Sicard, P.; Chaoui, H.; Cheriti, A.; Sechilariu, M.; Guerrero, J.M. A novel decoupled trigonometric saturated droop controller for power sharing in islanded low-voltage microgrids. *Electr. Power Syst. Res.* **2019**, *168*, 146–161. [CrossRef]
21. Fukuda, S.; Yoda, T. A novel current-tracking method for active filters based on a sinusoidal internal model [for PWM invertors]. *IEEE Trans. Ind. Appl.* **2001**, *37*, 888–895. [CrossRef]
22. Francis, B.A.; Wonham, W.M. The internal model principle for linear multivariable regulators. *Appl. Math. Optim.* **1975**, *2*, 170–194. [CrossRef]
23. Zmood, D.N.; Holmes, D.G. Stationary frame current regulation of PWM inverters with zero steady-state error. *IEEE Trans. Power Electron.* **2003**, *18*, 814–822. [CrossRef]
24. Teodorescu, R.; Blaabjerg, F.; Liserre, M.; Loh, P.C. Proportional-resonant controllers and filters for grid-connected voltage-source converters. *IEE Proc. Electr. Power Appl.* **2006**, *153*, 750–762. [CrossRef]
25. Pereira, L.F.A.; Flores, J.V.; Bonan, G.; Coutinho, D.F.; da Silva, J.M.G. Multiple resonant controllers for uninterruptible power supplies—A systematic robust control design approach. *IEEE Trans. Ind. Electron.* **2013**, *61*, 1528–1538. [CrossRef]
26. Callier, F.M.; Desoer, C.A. *Linear System Theory*; Springer Science & Business Media: Berlin, Germany, 2012.
27. Maccari, L.A.; Massing, J.R.; Schuch, L.; Rech, C.; Pinheiro, H.; Oliveira, R.C.L.F.; Montagner, V.F. LMI-Based Control for Grid-Connected Converters With LCL Filters Under Uncertain Parameters. *IEEE Trans. Power Electron.* **2014**, *29*, 3776–3785. [CrossRef]
28. Chilali, M.; Gahinet, P.; Apkarian, P. Robust pole placement in LMI regions. *IEEE Trans. Autom. Control* **1999**, *44*, 2257–2270. [CrossRef]
29. Gahinet, P. H∞ Design with Pole Placement Constraints: An LMI approach. *IEEE Trans. Autom. Control* **1996**, *45*, 358–367.

applied sciences

Article

Sizing and Sitting of DERs in Active Distribution Networks Incorporating Load Prevailing Uncertainties Using Probabilistic Approaches

Amin Foroughi Nematollahi [1], Hossein Shahinzadeh [1,*], Hamed Nafisi [1], Behrooz Vahidi [1], Yassine Amirat [2,*] and Mohamed Benbouzid [3]

1 Department of Electrical Engineering, Amirkabir University of Technology (Tehran Polytechnic), Tehran 15916-34311, Iran; amin.forooghi@aut.ac.ir (A.F.N.); nafisi@aut.ac.ir (H.N.); vahidi@aut.ac.ir (B.V.)
2 L@bISEN, ISEN Yncréa Ouest, Brest Campus, 20, Rue Cuirassé Bretagne, 29200 Brest, France
3 Institut de Recherche Dupuy de Lôme (UMR CNRS 6027 IRDL), University of Brest, 29238 Brest, France; Mohamed.Benbouzid@univ-brest.fr
* Correspondence: h.s.shahinzadeh@ieee.org (H.S.); yassine.amirat@isen-ouest.yncrea.fr (Y.A.)

Abstract: In this study, a microgrid scheme encompassing photovoltaic panels, an energy storage system, and a diesel generator as a backup supply source is designed, and the optimal placement for installation is suggested. The main purpose of this microgrid is to meet the intrinsic demand without being supplied by the upstream network. Thus, the main objective in the design of the microgrid is to minimize the operational cost of microgrid's sources subject to satisfy the loads by these sources. Therefore, the considered problem in this study is to determine the optimal size and placement for generation sources simultaneously for a microgrid with the objectives of minimization of cost of generation resources along with mitigation of power losses. In order to deal with uncertainties of PV generation and load forecasting, the lognormal distribution model and Gaussian process quantile regression (GPQR) approaches are employed. In order to solve the optimization problem, the lightning attachment procedure optimization (LAPO) and artificial bee colony (ABC) methods are employed, and the results are compared. The results imply the more effectiveness and priority of the LAPO approach in comparison with ABC in convergence speed and the accuracy of solution-finding.

Keywords: microgrid; optimization; lightning attachment procedure optimization (LAPO) algorithm; photovoltaic panel; uncertainty

Citation: Nematollahi, A.F.; Shahinzadeh, H.; Nafisi, H.; Vahidi, B.; Amirat, Y.; Benbouzid, M. Sizing and Sitting of DERs in Active Distribution Networks Incorporating Load Prevailing Uncertainties Using Probabilistic Approaches. *Appl. Sci.* **2021**, *11*, 4156. https://doi.org/10.3390/app11094156

Academic Editor: Amjad Anvari-Moghaddam

Received: 28 February 2021
Accepted: 29 April 2021
Published: 1 May 2021

Publisher's Note: MDPI stays neutral with regard to jurisdictional claims in published maps and institutional affiliations.

1. Introduction

In recent years, the increase in electrical demand, the rise of crude oil and natural gas prices, restructuring and the growth of privatization, and the advent of modern technologies have been led to revolutionary changes in the electricity industry and the assumption of specific attention to distributed generation (DG) technologies [1–3]. DG sources or small-scale electricity sources can generate power in the range of 1 kW to 10 MW in the location of the load or in the vicinity of consumption centers. DG technologies bring tangible benefits such as peak clipping (peak shaving), improvement of reliability indices, reduction in power losses due to being close to consumption places, alleviation of voltage drops and improvement of voltage profile, etc., for distribution networks. In addition, the integration of renewable clean energy sources, such as solar, wind, and fuel cell sources, has promoted the power system planners and experts to use these DG units as much as possible [4,5]. The increase in the pervasiveness of DGs and the combined use of various types of DG sources has resulted in the emergence of the microgrid concept. The microgrids are generally defined as small-scale power systems in distribution voltage level encompassing some DGs and some electrical and thermal loads usually accompanied by energy storage systems (ESS) [6,7].

In traditional power systems, the power generation was used to be centralized, and the power flow was always unidirectional from generation and transmission systems toward distribution systems and loads. However, in the recent decade, the structure of power systems has evolved so that modern power systems are experiencing more interest in the use of DG systems at the distribution level [8]. According to the rules and regulations in different countries, various definitions are given based on the place of DG installation, the purpose of use of DG, and their scale of generation (DG sizing). However, a common definition that overlaps all those definitions can be expressed as a small-scale generating unit with limited scalability, which is intended to be used in a distribution network or demand-side [9–16].

Determination of the optimal size of DG units and the optimal place for installation of them is distribution network subject to satisfy all constraints of the grid and minimize the incurred cost to the grid's operation have assumed particular attention since the last decade. In this regard, a wide diversity of studies is conducted on this subject, and various methods are proposed in order to solve and assess this problem. In [17], a microgrid scheme including a photovoltaic (PV) system is proposed, in which the microgrid works in grid-connected and isolated modes. In this work, the objective of the study is to minimize the power losses in the distribution grid. The authors in [18] have proposed an isolated microgrid scheme, which includes PV, wind turbine, diesel generator units, and the object is to minimize the overall cost. A hybrid system model is presented in [19], in which various types of DGs are used. However, as a critique, in this study, the places of DG units are supposed to be fixed, and no optimal placement for DGs has been done. In such a circumstance, by altering the installation location, the determined sizes are not valid as optimized scale anymore. Another research is carried out on the design of hybrid energy systems in [20]. The downside of this study is that the size, installation placement, and optimal performance are not considered into account simultaneously, and they are assessed separately. Such an evaluation cannot assure the optimal design and the optimal performance point of the hybrid system. Thus, all the optimization variables attributable to the design of hybrid systems and microgrids must be optimized at the same time rather than separately. Another critical factor is the consideration of uncertainties in PV generation and demand forecasting, which is not well addressed in the previous works. The authors in [21] have investigated a multi-objective DG placement and sizing problem subject to reduce loss and to enhance the voltage profile using the shuffled frog leap algorithm. Optimal sizing of renewable energy resources with the goal of loss reduction in distribution grids using the ant lion optimization method is presented in [22]. A battery placement problem is also presented in [23], in which it is objected to reducing losses in a distribution grid with high penetration of solar sources. In [24], optimal allocation and sizing of renewable distributed generation are investigated. Reference [25] has delved into the optimal sizing and placement of RESs in distribution systems considering load growth. The optimal allocation of DG units for a distribution network is presented in [26]. In addition, optimal sizing renewable and dispatchable DGs in distribution networks has been investigated in [27]. The authors in [28] have proposed a new method to deal with the optimization problem of optimal placement and sizing of energy storage systems subject to improve the reliability of hybrid power distribution networks encompassing renewable energy sources. In [29], a multi-objective dynamic and static reconfiguration model with the optimized placement of solar unit and battery storage system is proposed. A meta-heuristic algorithm is employed to find the optimal size and installation location of DG sources with the objective of loss reduction in [30,31]. As it is clear, none of the works in the literature has paid attention to the uncertainties of distribution-side resources. In other words, the uncertainties of load and demand-side generation sources can have a great impact on the operation and planning of the distribution sector as well as microgrids that should be taken into account. In [32,33], lightning attachment procedure optimization is employed to solve a non-smooth and non-convex dispatch problem, including uncertain variables pertaining to wind power sources. The authors in [34] also have presented a novel method to deal with the uncertainty of load forecasts as well as the uncertainty of renewable generation sources to solve a placement problem. Similar studies are conducted in which the uncertainty of load, price, and renewable sources are investigated [35–37].

In this paper, a hybrid energy system is designed, which is regarded as a microgrid in a distribution network. The costs of operation, maintenance, and investment are also taken into account. In order to deal with uncertain variables of solar generation, the lognormal distribution model is employed to exploit the trends and patterns of solar irradiance. In addition, the Gaussian process quantile regression is applied as the forecasting approach to deal with uncertainties of load forecasting. A novel optimization algorithm is employed in order to optimize the performance and to find the optimal location for the installation of DGs. It is supposed that the microgrid is connected to the upstream network. Hence, the cost of energy exchange with the main grid is also taken into consideration in the objective function. The proposed scheme is implemented on the 69-bus IEEE test system in order to evaluate the effectiveness of the model. In Section 2, the problem outlines are described. In this section, the objective function and the constraints of the problem are expressed. In Section 3, the employed optimization algorithm is explained, and in Section 4, the simulation and results are discussed. Ultimately, in the last section, the conclusions are drawn.

2. Problem Outlines

As declared, the purpose of this study is to design a microgrid scheme subject to supply the loads in a distribution feeder or network. This microgrid consists of a PV panel, an energy storage system, and a diesel generator as a backup source of energy. The primary goal of this study is to determine the optimal size of the PV panel, diesel generator, and the sufficient quantity of batteries to maintain an uninterrupted supply for the loads with respect to the objective function of the problem. Hence, at first, the demand for the microgrid must be specified.

2.1. Microgrid's Demand

The contemplated microgrid is a local distribution network that has a variable load during a day. With regard to the incremental trend of consumption of residential and industrial loads at the early hours of the evening, the peak of the demand curve is supposed to be in the evening. In addition, with regard to the consumption pattern of industrial loads, another peak exists around midday hours. Thus, in order to consider the load pattern of the microgrid, a 24 h load profile with two peaks is taken into account as figured out as follows. Figure 1 demonstrates the normalized demand of the microgrid. It is noteworthy to assert that the load of the distribution network is also determined in a similar way.

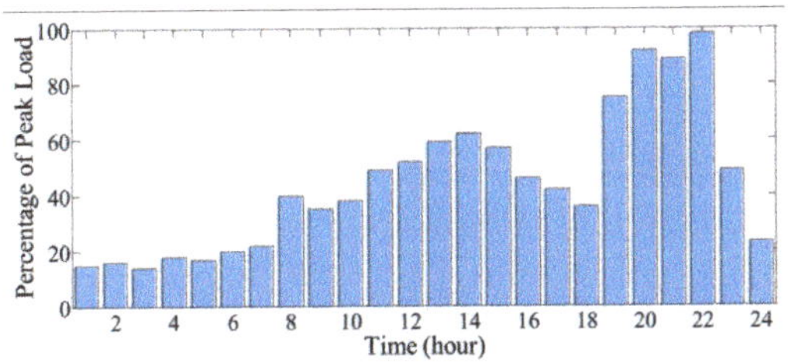

Figure 1. The normalized demand curve of the microgrid.

2.2. Microgrid's Supply Strategy

The first goal of size determination of microgrid's generation sources is to supply the loads in this micro-scale grid. The PV panels have a vital role in power provision for the loads. However, when PV panels cannot generate sufficient power, the energy storage unit and diesel generator unit are responsible for meeting the loads. Thus, the optimal supply strategy can be express as follows:

2.2.1. Enjoying Photovoltaic Panel's Power

A. If the PV generation is more than the grid's demand and the batteries are not fully charged, the excess PV generation should be stored in the batteries;
B. If the PV generation is more than the grid's demand and the batteries are fully charged, the excess generation of PV panels must be sold to the distribution network;
C. If the PV generation is lower than the grid's demand, a portion of the loads will be supplied through PV generation, and the rest of the demand must be procured through the energy storage system;
D. If the PV generation is lower than the grid's demand and the battery system is not adequate to supply the loads entirely, the loads will be satisfied by PV panels and energy storage units as much as possible, and the rest of the loads must be met by the diesel generator;
E. If the total generation by PV panels, batteries, and the diesel generator is not sufficient for satisfying the load, the load-shedding measure must be imposed for the excessive demand.

2.2.2. Being Deprived of PV Generation

A. If the batteries' capacity is adequate to supply the demand, the loads will be served by the batteries solely;
B. If the batteries' capacity is not sufficient to supply the demand, the rest of the loads must be served by the diesel generator;
C. If the demand is more than the combined capacity of PV panels and the diesel generator, the rest of the loads must be remained unsupplied by load shedding.

It should be noted that the unsupplied load cannot be higher than 25% of the total demand of the grid. In addition, the capacity of the diesel generator must be so determined that it is not allowed to operate less than 20% of its maximum capable generation.

2.3. *The Objective Function*

When the suggested capacities for the generation sources are adequate to supply the microgrid's loads properly, the excess generated power can be sold to the distribution network. However, it must be determined which bus in the distribution network must be chosen, from which the excess generation will be injected into the distribution grid. The injection point has a considerable impact on the performance of the distribution grid. Thus, the location of the microgrid is an important item. Hence, the optimal placement of the microgrid must be investigated. The objective function for the optimal placement of the microgrid can be expressed as follows:

$$\min f = C_{PV} + C_{loss} + C_{bat} + C_{dg} - B_{ex} \tag{1}$$

2.3.1. Power Losses

In order to model the power losses, the losses are supposed to be as a power, which must be bought from the upstream network. Thus, it is modeled as a cost in the overall objective function equation. This objective can be expressed as follows:

$$C_{loss} = \sum_{d=1}^{365} \sum_{t=1}^{24} \sum_{i=1}^{N_b} P_{loss}^{i,t,d} \times \rho^{t,d} \tag{2}$$

In the above equation, $\rho^{t,d}$ denotes the price of electricity on the day of d at hour t, and $P_{loss}^{i,t,d}$ stands for the number of losses on the day of d at hour t in the ith line. It is noticeable that the price of electricity at peak, mid-peak, and off-peak hours is different.

2.3.2. The Cost Related to the Photovoltaic System

Deployment of solar energy incurs installation cost of PV panels, operation and maintenance cost, and replacement cost. Thus, the cost pertaining to PV panels can be modeled as below. The subscript of *PV* a is a symbol for a photovoltaic unit, *O&M* addresses operation and maintenance cost, *inv* indicates investment cost, and *rep* corresponds with replacement cost.

$$C_{PV} = C_{PV\_inv} + C_{PV\_O\&M} + C_{PV\_rep} \tag{3}$$

2.3.3. The Battery Cost

The energy storage facility in this work is a set of batteries that have similar costs, such as the PV panel, and can be formulated as Equation (4). The subscript of *bat* stands for battery energy storage system, *O&M* addresses operation and maintenance cost, *inv* indicates investment cost, and *rep* corresponds with replacement cost.

$$C_{bat} = C_{bat\_inv} + C_{bat\_O\&M} + C_{bat\_rep} \tag{4}$$

2.3.4. The Cost Pertaining to Diesel Generator

The cost of such equipment consists of investment cost, maintenance cost, and replacement cost. The subscript of *dg* stands for the diesel generator, *O&M* addresses operation and maintenance cost, *inv* indicates investment cost, and *rep* corresponds with replacement cost.

$$C_{dg} = C_{dg\_inv} + C_{dg\_O\&M} + C_{dg\_rep} \tag{5}$$

2.3.5. The Cost Pertaining to Load Shedding

If the load-shedding measure has to be imposed, this matter incurs a cost to the model, which must be included in the objective function of the problem.

2.3.6. The Profit Yielded by Excess Generation Selling

If the power generated by the PV panel is higher than the demand of the microgrid and charging capability of batteries, the excess power can be sold to the distribution network and earn a profit. The obtainable benefit (profit) can be calculated by the following equation:

$$B_{ex} = \sum_{d=1}^{365}\sum_{t=1}^{24} P_{ex}^{t,d} \times \rho^{t,d} \tag{6}$$

In above, B_{ex} is the benefit (profit) gain by the sale of surplus generation, and $P_{ex}^{t,d}$ shows the amount of surplus power on the day of d and at hour t.

2.3.7. Constraints

The optimization problem is accompanied by a set of technical constraints that restrict the solution space and must be considered in the model. These constraints can be defined as follows:

The power balance equality constraint during all intervals of a day:

$$\sum_{t=1}^{24} P_{PV}^{t} + P_{dg}^{t} + P_{bat}^{t} - P_{D\_mic}^{t} = 0 \tag{7}$$

The maximum permissible load shedding in the microgrid:

$$\sum_{t=1}^{24} P_{sh}^{t} \leq 0.25 \times \sum_{t=1}^{24} P_{D\_mic}^{t} \tag{8}$$

Generation balance in the distribution network must be observed.

$$\sum_{t=1}^{24} P_{g}^{t} + P_{ex}^{t} + P_{slack}^{t} - P_{loss}^{t} - P_{D\_dis}^{t} = 0 \tag{9}$$

The boundaries of the photovoltaic panel and diesel generator must be met.

$$0 < P_{PV} < P_{PV}^{\max} \tag{10}$$

$$0 < P_{dg} < P_{dg}^{\max} \tag{11}$$

The voltage of buses must be restricted within their permissible range.

$$V_{\min} < V < V_{\max} \tag{12}$$

The power flow passing through the lines must be restricted.

$$F_b < Limit_b \tag{13}$$

In the above equations, P_{PV}^t, P_{dg}^t, and P_{bat}^t represent power generation of PV panel, diesel generator, and batteries, respectively. Moreover, $P_{D\_mic}^t$ indicates the consumption of the microgrid at hour t. The shed load of the microgrid is shown by P_{sh}^t. In addition, P_g^t, P_{ex}^t, P_{slack}^t, P_{loss}^t, and $P_{D\_dis}^t$ represent the power generated by the generators in the distribution network, the exchanged power of the microgrid with the distribution network, the power maintains from the slack bus of the distribution network, the power losses in the distribution network, and the consumption of the distribution network at hour t. Moreover, P_{PV}^t, P_{dg}^t, V, V_{min}, V_{max}, F_b, and $Limit_b$ stand for the maximum capable power generation by PV panel, the maximum capable power generation by a diesel generator, bus voltage, the minimum and maximum permissible voltage of buses, the flow passing through branch b, and the thermal limit of branch b.

The prevailing constraints of the problem can be divided into two general categories of network constraints and storage constraints that must always be observed when seeking for optimal point in the solution space. Network constraints include the following two constraints:

As the thermal limit constraints express, the feeder power flow is not allowed to violate a specific cap.

$$S_k \leq \text{limit}_k \tag{14}$$

The hourly constraints pertaining to the energy storage unit consist of two constraints of the maximum storage capacity and charge/discharge rate limits. This limitation expresses that the storage charge level must always be below the maximum storage capacity and can be formulated as follows:

$$P_{min\_ch} \times I_{ch}(t) \leq P_{ch}(t) \leq P_{max\_ch} \times I_{ch}(t) \tag{15}$$

$$P_{min\_dch} \times I_{dch}(t) \leq P_{dch}(t) \leq P_{max\_dch} \times I_{dch}(t) \tag{16}$$

$$SoC_{min} \leq SoC(t) \leq SoC_{max} \tag{17}$$

$$SoC(t+1) = SoC(t) - P_{dch}(t) \times I_{dch}(t) + P_{ch}(t) \times I_{ch}(t) \tag{18}$$

$$I_{ch}(t) + I_{sch}(t) \leq 1 \tag{19}$$

In this equation, t represents the time, P_{ch} and P_{dch} denote the charge and discharge rates, and SoC stands for the state of charge of the storage unit that has a maximum storage capacity limit and a minimum storage capacity boundary owing to the long-run operation and maintenance facets. Each storage unit can be charged or discharged by a limited amount within an hour. The last equation also ensures that the storage unit cannot operate at charge and discharging modes simultaneously. In this equation, $I(t)$ is a binary variable.

2.3.8. The Uncertainty Modeling for PV Sources

The nature of solar power and photovoltaic irradiance can be represented by various types of distribution functions. One of the most efficient types for this purpose is the lognormal probability distribution function (PDF), which effectively and precisely characterizes the intensity of irradiance on a typical day. The PDF of solar irradiance (ζ_s) trailing the lognormal distribution function with a mean value of μ_s and standard deviation of σ_s that is presented in Equation (20).

$$f\left(\zeta_s \middle| \mu_s, \sigma_s^2\right) = \frac{1}{\sigma_s^2\sqrt{2\pi}} \exp\left\{-\left(\frac{(\log(\zeta_s) - \mu_s)^2}{2\sigma_s^2}\right)\right\} \quad \forall \zeta_s > 0 \tag{20}$$

The solar power generation (P_{PV}) depends on the solar radiation intensity, also known as solar irradiance, which is shown by ζ_s. The generation curve can be different based on different items such as the location and type of installation, the technology of solar panels, as well as the ambient temperature. A typical photovoltaic irradiance-power curve is given in Equation (21).

$$P_{PV}(\zeta_s) = A_{PV}\eta_{pv}\zeta_s \tag{21}$$

$$\eta_{pv} = \eta_0[1 - 0.0042\,(\frac{\zeta_s}{18} + T_a - 20)]\,\eta_{\text{inv}} \tag{22}$$

The amount of power generation by a solar panel depends on different factors, which can be estimated by Equations (21) and (22). Above, η stands for efficiency, and T_a denotes the ambient temperature.

In order to model the intermittencies, a dataset of historical records associated with the solar radiation. Such irradiance datasets can effectively be matched with a lognormal or Weibull distribution function. These distribution models usually have high mathematical compatibility with various natural phenomena. The PDF helps to generate randomly derived samples within the occurrence range while specifying a confidence range to eliminate low-probable cases. For each scenario, a random value is derived from the lognormal model. If the number of scenarios is high, various scenario reduction techniques can be employed that merge similar cases into one class. Based on the model objectives, the operation state can be adjusted within the confidence range with respect to the level of risk chosen by the operator. The operator can arrange the operation point in a risk-averse zone to ensure a reliable and secure operation, although it incurs higher costs. On the other hand, the operation of the grid, with respect to the whole uncertainties, can be stated in a high-risk zone by ignoring less-probable incidences to boost the profitability and the economy of the model.

2.3.9. Load Forecast Uncertainty

Load forecasting is a critical item in many planning applications in power systems. Load forecasting can also be performed in different intervals for various purposes. Load forecasts are dependent on weather conditions (temperature, humidity, air brightness, wind speed). In addition, each day of the week has its own load curve. Load consumption curves on holidays and non-holidays are also different from each other. In different seasons of the year, according to the factors specific to each season, such as day length, the load consumption pattern will be different. Short-term load forecasting is a pivotal data needed for running a day-ahead market. The results of STLF are employed by GENCOs to discover prices with regard to the way other market participants may act. Price forecast signals are also applied to bidding strategy techniques to maximize profitability. The system operator will clear the market for the next-day delivery according to the forecasted values along with the bids from GENCOs. ISO is responsible for balancing the market between generation level and the forecasted demand at each interval. However, with regard to the inaccuracy of forecasts, there is always a violation that must be redressed by storage units or other controlling actions such as load shedding.

There are multiple approaches to deal with load forecasting. The most common methods are the time series method, regression methods, and intelligent methods. Intelligent methods are also classified into several categories such as artificial neural networks, fuzzy logic-based methods, also ANFIS models, wavelet transformation methods, as well as support vector machines. In order to deal with uncertainties, probabilistic models can be integrated into the employed load forecasting paradigm. In this study, the Gaussian process quantile regression (GPQR) is employed to deal with uncertainties [38]. Due to the stochastic nature of load patterns as well as various external factors such as weather conditions, calendar effects, and seasonal factors, the power demand signals exhibit intermittent and volatile characteristics. Hence, a prediction scheme is needed to provide the most probable distribution of power demand patterns rather than a crisp value. A Gaussian process model is a set of random variables, in which any member corresponds with a probability distribution that can be represented by the following function:

$$f(l) \sim GP(\mu(l), COV(l,l')) \tag{23}$$

In this equation, $\mu(l)$ and COV(l,l') denote the mean and covariance functions that can be calculated as follows:

$$\mu(l) = E\{f(l)\} \tag{24}$$

$$COV(l,l') = E\left\{[f(l) - \mu(l)] \times [f(l') - \mu(l')]^T\right\} \tag{25}$$

The covariance function illustrates the similarity between data points. One of the most widely-used functions for describing covariance is the squared exponential (SE) model that can be represented as follows:

$$COV_{SE}(l,l') = \theta_f^2 \exp\left(\frac{\|l - l'\|}{\theta_{length}^2}\right) \tag{26}$$

The parameters θ_f and θ_l control the scale and length. With regard to the differentiability of this covariance, which implies that the Gaussian process is very smooth, a simplified alternative, known as Matern covariance, can be replaced because, in practice, no phenomena do not reflect such a strong smoothness.

$$COV_{Mat}(l - l') = \sigma^2 \frac{2^{1-v}}{\Gamma(v)} \left(\sqrt{2v}\frac{l - l'}{i}\right)^v B_v\left(\sqrt{2v}\frac{l - l'}{i}\right) \tag{27}$$

In the above, B_v represents the modified Bessel function, in which v and i are both positive hyperparameters. To simplify this covariance function, the value of v is supposed to be $v = p + 1/2$. In this parameter, p is a non-negative integer. The value of v is set as 5/2 or 3/2 in most previously conducted research studies, which subsequently and, respectively, are named Mat5 and Mat3. The probabilistic prediction model also has another covariance function named the period covariance, which is useful to model periodic phenomena and can be described as follows:

$$COV_{Mat}(l,l') = \theta_f^2 \exp\left(-\frac{2}{\theta_{length}^2} \sin^2\left(\pi\frac{l - l'}{p}\right)\right) \tag{28}$$

Typically, the load forecasts are highly influenced by a variety of features that can be defined by Equation (29). According to this equation, load at time t depends on the similar hours ($t \in \{0, 24\}$) throughout the historical records, the day of the year ($d \in \{1, \ldots, 365\}$), the value of the load at similar intervals, the value of weather variables such as temperature, and the price at similar intervals.

$$\hat{y} = f(t, d, v_l, v_w, price) \tag{29}$$

GPQR method seeks for the relationships and correlation between inputs and output based on a probabilistic framework. Quantile regression (QR) delineates a type of regression analysis that detects and exploits the relationships between quantiles of the conditional distribution of a response variable and input variables. The least absolute deviations regression integrates the median of the conditional distribution, which is called the norm regression case of quantile regression shown by L_1. Unlike least-squares regression and norm regression, the quantile regression encompasses minimization of the summation of asymmetrically weighted absolute residuals. Hence, QR can find more sophisticated relationships between inputs and outputs with better precision. QR has more flexibility and compatibility to deal with large datasets, such as market analysis or econometrics. With regard to the accuracy of predictions, the loss function can be described as follows:

$$L_\tau(\varepsilon_i) = \begin{cases} \tau\varepsilon_i & if \quad \varepsilon_i \geq 0 \\ (\tau-1)\varepsilon_i & if \quad \varepsilon_i \geq 0 \end{cases} \tag{30}$$

The required quantile is defined by $\tau\in[0, 1]$ and $\varepsilon_i = y_i - \hat{y}_i$ so that y_i is the actual model and $\hat{y}_i$ denotes the predicted quantile model. So far, different linear programming methods are employed to achieve the desired quantiles through direct loss function minimization, which leads to the maximization of a likelihood. To solve this drawback, the Gaussian process is incorporated into the QR model. The density function of loss based on this model can be represented as follows:

$$L(t|\mu,\sigma,\tau) = \frac{\tau(1-\tau)}{\sigma}\exp\left[-\frac{t-\mu}{\sigma}(t - I(t \leq \mu))\right] \tag{31}$$

where $\tau\in[0, 1]$ is responsible for shaping and controlling the skewness of the distribution curve, μ stands for the mean value, and σ denotes the standard deviation, which should always be positive. The binary variable I takes the value of 1 when the condition is true; otherwise, it takes 0.

$$U_\tau(y,q) = Z\exp\left[-\sum_{i=1}^{N} L_\tau(y_i,q_i)\right] \tag{32}$$

In Equation (32), q denotes the predicted value of the τ quantile, Z is the normalization constant. Afterward, a Gaussian process is placed on the QR function:

$$p(q) = GP(q|0,K) \tag{33}$$

The GPQR model training can be conducted by integral maximization. The expectation propagation algorithm can be employed to locally approximate this integral.

$$\text{argmax}\int_q U_\tau(y,q)p(q)\ d(q) \tag{34}$$

Suppose a dataset of historical load records as $l = \{x_1, x_2, \ldots, x_N\}$, which are independently distributed samples. The estimated shape of this distribution function can be obtained through a Gaussian kernel density estimator model.

$$K(x_1,x_2,\sigma) = \frac{1}{\sqrt{2\pi}\sigma}e^{-\frac{(x_1-x_2)^2}{2\sigma^2}} \tag{35}$$

Mean absolute percentage errors (MAPE) and root mean squared error (RMSE) are two employed evaluation metrics that can exhibit the performance of the forecasting model.

$$RMSE(y_i,\hat{y}_i) = \sqrt{\frac{1}{N}\sum_{i=1}^{N}(y_i,\hat{y}_i)^2} \tag{36}$$

$$MAPE(y_i,\hat{y}_i) = \frac{1}{N}\sum_{i=1}^{N}\left|\frac{y_i,\hat{y}_i}{y_i}\right| \tag{37}$$

2.4. Energy Management Paradigm by Microgrid's Controller

The controlling framework of a microgrid is a vital element in this planning scheme. The microgrid controller, also known as the microgrid operator, is responsible for collecting raw data and analyzing them to find the optimum operation point with respect to all operational constraints of both sides. In other words, it has to provide an accurate estimation of generation sources for all intervals and take a sensible range of risk for being preserved from the detrimental zone. Then it can dedicate the level of power exchange with the upstream grid for various hours. To implement such an intelligent autonomous mechanism, a smart environment is needed that is only feasible through IoT infrastructures [39]. The following flowchart concisely describes the proposed scheme. The paradigm of the proposed scheme is depicted in Figure 2.

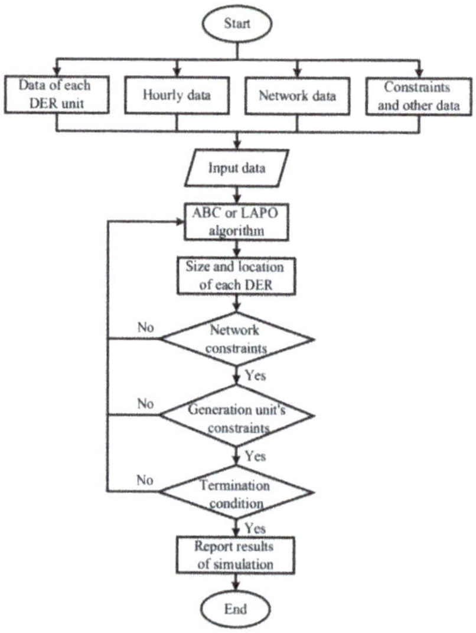

Figure 2. The paradigm of the proposed scheme.

3. Lightning Attachment Procedure Optimization (LAPO) Method

This algorithm was inspired by the procedure of lightning the attachment process to the ground or any type of earthed object. In order to simulate this optimization algorithm, at first, some test points are considered on a cloud or the ground as the initial population. By increasing the electric field at the points on the cloud, the streamer channels begin to break down the air and move toward the ground, which is called a downward leader. As these downward leaders move toward the ground, the opposite charges will be enhanced in the ground and produce upward leaders. The striking point at the final jump step is a point where the upward leader collides with the downward leader [40–42]. The algorithm is composed of some sections that will be presented as follows:

3.1. Initialization

At first, the test points on the surface of a cloud as well as on the ground are defined. These dedicated charges are called the initial population.

3.2. The Next Jumping Node

Each branch of the lightning has some probable points in front of itself, which can move toward one of them. The choice of the next jumping point of the lightning is highly correlated with the intensity of the electric field between the point and the probable points. Of course, the next jump will not necessarily be toward the point with the maximum field, and part of the process of selecting the next point will occur randomly. For the mathematical modeling of this section, a random node is selected for each node in the search space. If the electric field in this point (merit value of the point or the fitness value at the intended point) is greater than the background field (which is supposed to be the mean field), the lightning moves toward this point; otherwise, the path of lightning will be in another direction. It is noticeable that the term opposite direction does not imply an upward movement exclusively rather than an angled motion toward this point. Hence, for determining the next jump, the following equation can be employed:

$$P = sign(F_{av} - F(r)) \tag{38}$$

$$X_{new}(i,j) = X(i,j) - P \times rand \times (Xav(1,j) - X(r,j)) \tag{39}$$

3.3. Streaming Forward and Elimination of Branches

When a new lightning branch is formed, this branch can stream downward until the branch's charge cuts down and becomes lower than a specific value (critical value). The critical value is assigned to be 1 μC because the air cannot be broken down for the lower levels of charges, and the branch will be faded.

3.4. Upward Leader Propagation

When the branches start to move downward, the upward leaders start to move upward. The motion speed of the upward leader is correlated with the amount of charge aggregated at this point as well as the total charge of the downward branch. The distribution of charge in the downward branch is so that there are fewer charges near the cloud and more charge at the tip of the leader. This is why the tip of the leader is brighter. This incremental charge aggregation over the branch can be considered as a linear or exponential function. Here, the branch's charge is considered exponentially, as pointed out by the following two equations:

$$S = 1 - (t/t_{\max}) \times \exp(-t/t_{\max}) \tag{40}$$

$$Cc = S \times (X_{\min}(i,j) - X_{\max}(i,j)) \tag{41}$$

Above, C_C denotes the branch's charge, $X_{\min}$ is the charge of the branch in the tip of the leader, $X_{\max}$ indicates the branch's charge at the beginning point. The upward leaders will be propagated according to the following equation:

$$X_{new}(i,j) = X(i,j) + rand \times Cc \tag{42}$$

3.5. Convergence

When an upward leader reaches a downward leader, the collision point of the lightning will be determined. In other words, when the optimum point is specified, the algorithm will be terminated. The abovementioned procedure will be iteratively executed until the optimum point is found.

4. Simulation and Results

In this section, the optimal placement for the microgrid is carried out on the targeted test system. The targeted system is a radial 69-bus IEEE test system, which is depicted in Figure 3 in the form of a single-line diagram. In addition, in order to have a better appraisal of the effectiveness of the proposed method, the proposed method is also tested on a radial 33-bus IEEE test system, which is depicted in Figure 4. However, the input data is the same for both grids. In this paper, all the methods are implemented in MATLAB 2018a in a Core i5 PC with 3 GHz processing frequency of CPU and 8 GB of RAM. The convergence behavior is another aspect with which the methods are compared to each other.

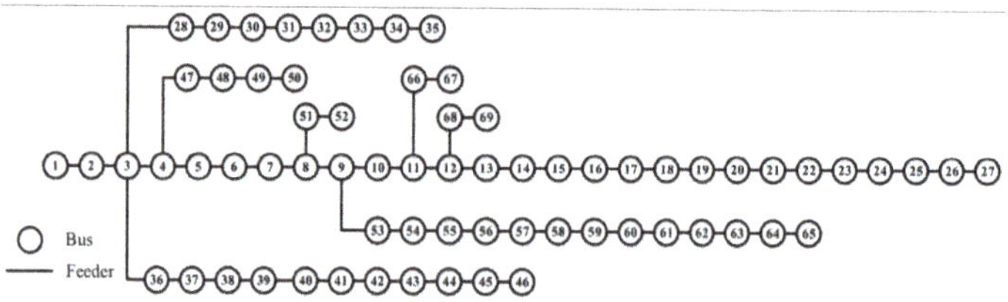

Figure 3. The single-line diagram of the 69-bus test system.

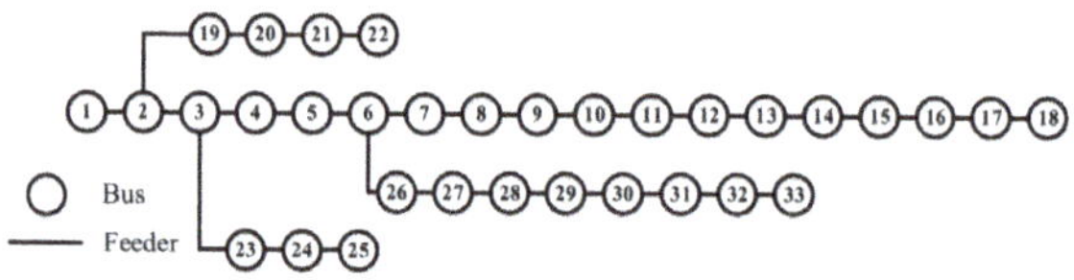

Figure 4. The single-line diagram of the 33-bus test system.

The characteristics corresponded with the lines and the buses of this system refer to [43,44], respectively. The normalized demand of the microgrid is shown in Figure 5. The demand for this microgrid at the peak is considered to be 250 kW. The peak of the load curve of the grid is supposed to be the mid-peak demand obtained from [43,44]. Moreover, the hourly solar insulation profile is normalized and illustrated in Figure 6.

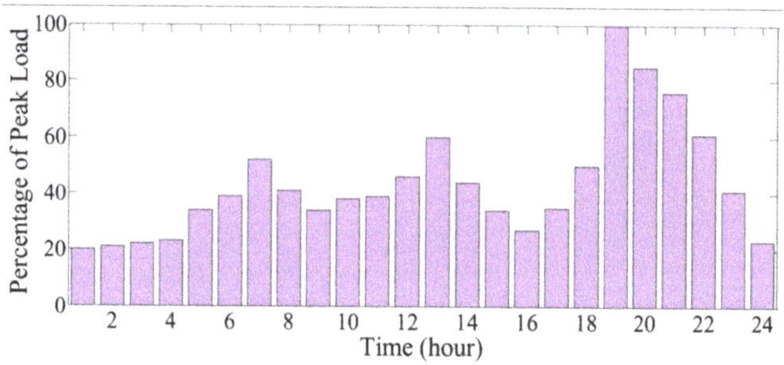

Figure 5. The normalized demand of distribution network for both systems.

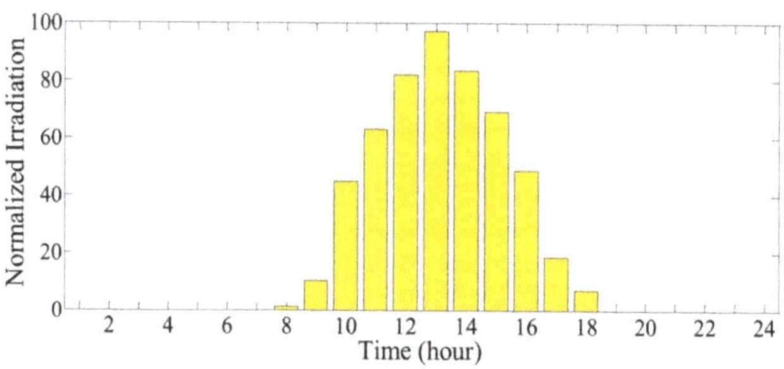

Figure 6. The normalized daily solar irradiance for both distribution systems.

The maximum generation capability of the PV panel per month is shown in Table 1. The placement mechanism is executed in the worst condition. Thus, the calculations are conducted for the month of August. This month indicates the lowest level of mean solar radiation.

Table 1. The mean solar radiation for each month for both radiation systems. [45].

Jan.	Feb.	Mar.	Apr.	May	Jun.	Jul.	Aug.	Sep.	Oct.	Nov.	Dec.
5.45	5.73	6	6.01	5.65	5.26	4.81	4.78	5.15	5.63	5.84	5.57

The economic data for different sources were extracted from reference [44]. The investment cost of the PV panel is considered to be 7.44 €/kW, and the O&M cost is supposed to be 40 €/year. The cost of each battery is determined as 0.283 €/Ah, and each battery incurs 50 €/year for O&M costs. The operation cost and maintenance cost of the diesel generator is 0.55 €/W and 0.2 €/h, respectively.

The demand in the targeted microgrid is supposed to be composed of residential and industrial loads. Table 2 delineates the price of electricity corresponded with the types of loads and the demand level.

Table 2. The electricity price for different load types and various demand levels for both distribution systems.

Load Level	Price of Residential Load (€/MW)	Price of Industrial Load (€/MW)	Percent of Peak Load
Low	35	2000	50 < L
Middle	49	2800	50 < L < 70
Peak	70	3050	L > 70

4.1. *The IEEE Test Systems (33-Bus and 69-Bus)*

The optimization for the 33-bus test network is conducted with regard to the acquired economic data. The results are demonstrated in Table 3. In addition, the convergence behavior of two targeted methods in reaching the optimal solution is illustrated in Figure 7. Accordingly, the LAPO method has reached the optimal solution with fewer iteration. In terms of accuracy, the results of LAPO are better than the ABC algorithm. As can be seen, at the early hours of the time horizon (at midnight), when there is no solar power generation, a battery energy storage unit can supply the demand of internal loads of the microgrid as it is clear that the battery power is negative during these hours, which indicates that it is discharged. The battery starts charging after 8:00 and reaches its maximum charge level at 11:00. Therefore, no power is stored in it until it starts discharging again at 18:00. It is recharged at 19:00 and fully discharged at 20:00 due to the high volume of consumption. The diesel generator commits at 8:00 and stops power exchange at 9:00. These conditions continue until 20:00, and after that, it continues to serve the loads until midnight. Due to the significant amount of solar production, a large volume of this production is sold to the distribution network, and the obtained profit improves the objective function. As it is evident, the network is designed in such a way that no load is interrupted during the day and all the loads are completely satisfied.

Table 3. The design and size determination of microgrid for the distribution test systems (33-bus and 69-bus).

Bus	Method	Location	PV Size (Kw)	Diesel Size (Kw)	Number of Batteries	Cost (€)
33	LAPO	6	395	242	25	47,162
	ABC	26	393	242	25	47,383
69	LAPO	59	400	256	26	50,360
	ABC	61	399	238	25	52,351

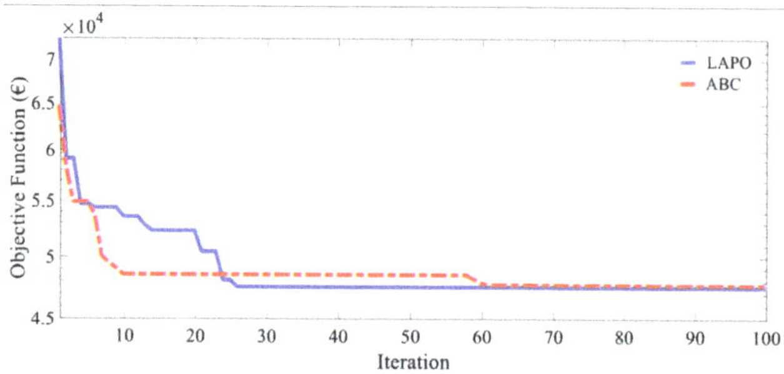

Figure 7. The comparison of the convergence behavior of different methods in seeking for the best solution for the 33-bus distribution test system.

With regard to the economic data provided in the previous section, the optimization problem is simulated on the 69-bus test system. The optimized results of the simulation are presented in Table 3. In addition, the convergence behavior of both optimization techniques employed in this study is depicted in Figure 8. As can be seen, it is evident that the LAPO approach not only has reached the optimal solution in less iteration, but also it has found a better solution in comparison with the ABC method.

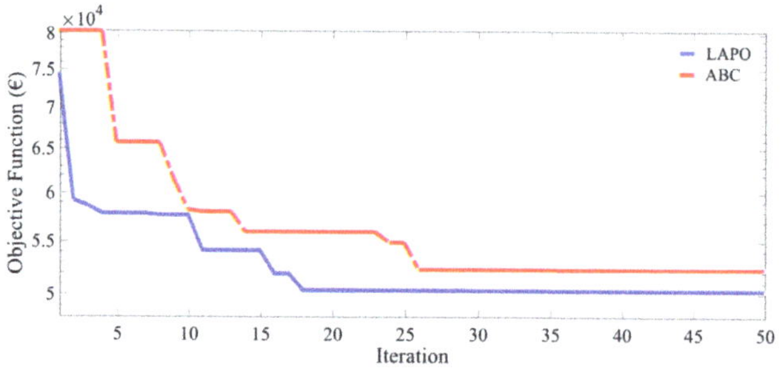

Figure 8. The comparison of the convergence behavior of different methods in seeking for the best solution for the 69-bus distribution test system.

The behavior of the microgrid in August is illustrated in Figure 9. As can be perceived, in the early hours at midnight, while the sunrise has not occurred, the battery is responsible for maintaining power for the microgrid. Thus, the power battery is negative in these hours, which implies that the battery is discharging. The battery starts to be charged at 8:00 in the morning, and it will be fully charged at 11:00. Hence, the batteries will not be charged anymore and remain idle until 18:00. At 18:00, the batteries are to some extent discharged. However, they are charged again at 19:00. With regard to the high level of consumption at 20:00, the batteries deliver all of their stored energy to the microgrid. The diesel generator is connected to the microgrid at 8:00, and it is disconnected at 9:00. It continues until 19:00. Again, at 19:00, the diesel generator starts to supply the load until the end of the time horizon of this study (24:00). With respect to the high level of solar generation, a considerable share of this production is sold to the distribution grid. The obtained economic benefit has improved the objective function. As it is clear, the microgrid is so designed that the loads are supplied entirely and load shedding has not occurred. The microgrid behavior in August is depicted in Figure 9.

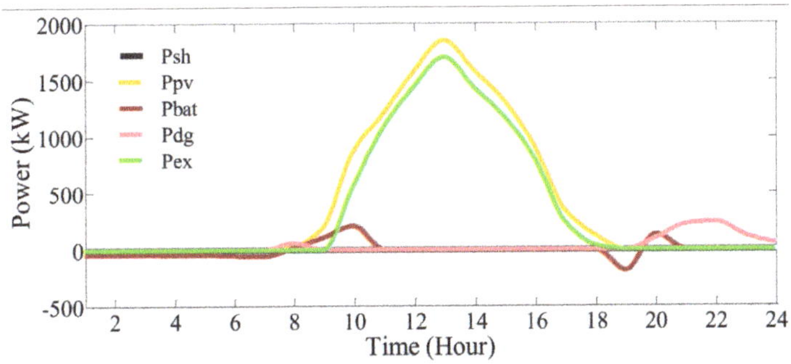

Figure 9. The behavior of the microgrid in August.

The power purchased from the main grid is shown in Figure 7. As it shows, during the day hours when there is solar energy, the demanded power from the upstream grid is drastically reduced. It should also be noted that the network losses and the corresponding costs were 1061 kW and 55.86 € per day, which reached the values of 956.79 kW and 51.78 € per day, after the installation of microgrids in the network, respectively. Figure 8 shows the amount of load shedding in the distribution network at different times of the day. The results figure out that during the hours when there is solar energy, solar unit injects power into the distribution network, the amount of shed load has reached zero, which is very beneficial in terms of cost and efficiency for the distribution network as well as reliability improvement from microgrid point of view. The model is also tested on the 69-bus test system. The results imply that the bought power from the main grid by the 69-bus distribution network is illustrated in Figure 10. As it is evident, during the hours that the PV panel can generate power, the absorbed power from the sub-transmission grid is considerably diminished. The power losses of the distribution grid and the subsequent cost of them were about 1350 kW equivalent with 65.76 €/day before the implementation of the microgrid scheme, and it is alleviated to 989.32 and 56.28 €/day after implementation of the microgrid. In Figure 11, the hourly shed load in the distribution network is shown. It is obvious that whenever the PV generation exists, all of the loads in the distribution grid are satisfied, and no load-shedding measure in the distribution network is executed. This matter significantly improves the performance of the distribution network.

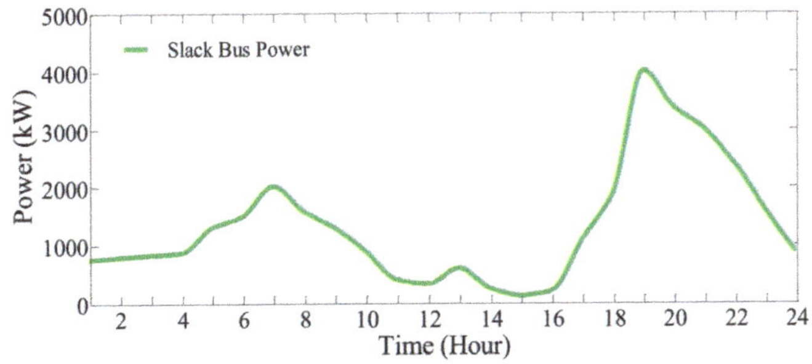

Figure 10. The bought power from the main power system by the distribution network.

4.2. Incorporation of Uncertainty in the Model

The demand for a grid and the daily radiation intensity are not deterministic parameters, and they usually have uncertainties. Hence, these parameters must be forecasted, and the scheduling must be conducted based on the forecasts. Therefore, the existence of forecasting error is inevitable, and the error must be included in the optimization model in order to mitigate the risk. A common way to model the uncertainties corresponded with solar radiation and demand of the grid is to employ a normal probability distribution. In other words, the mean value and the standard deviation of these parameters are calculated, and the design and the scheduling must be carried out based on these forecasts.

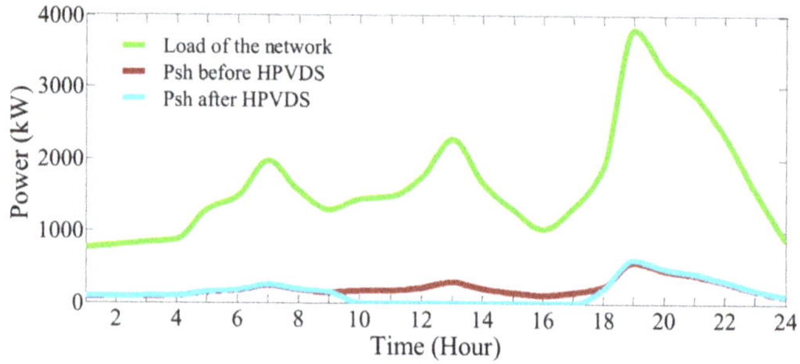

Figure 11. The amount of shed load in the distribution network in the presence and absence of microgrid.

The most famous and the most accurate approach for dealing with uncertainties and probabilistic problems is the Monte Carlo algorithm. In this approach, a large number of probabilistic samples are defined within a specified range, and the scheduling is performed for all of these samples. Then, the probability density function of targeted parameters is extracted. In this study, an upper and a lower boundary are dedicated for the demand of microgrid and distribution network along with the daily solar radiation. It is supposed that the radiation and demand will be materialized within the defined range. These boundaries are depicted in Figures 12–14.

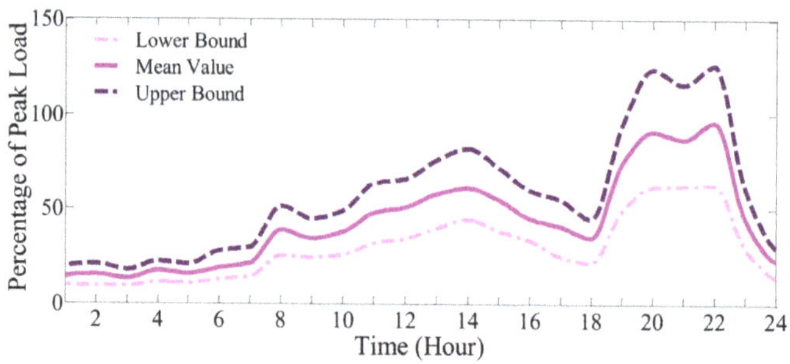

Figure 12. The upper bound, the lower bound, and the mean value of demand of microgrid in the probabilistic study.

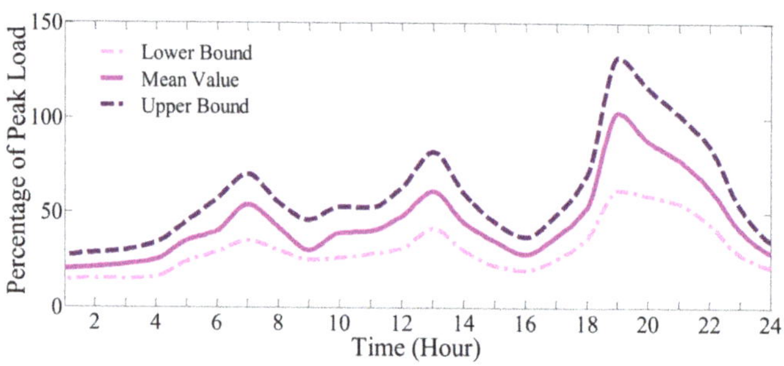

Figure 13. The upper bound, the lower bound, and the mean value of demand of distribution network in the probabilistic study.

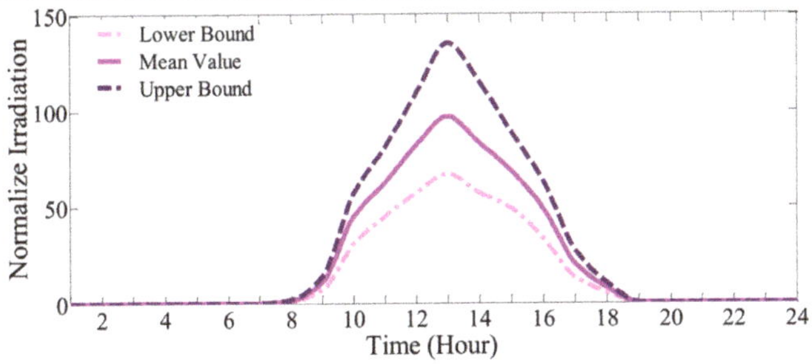

Figure 14. The upper bound, the lower bound, and the mean value of daily solar irradiance in the probabilistic study.

In order to solve the stochastic problem, the Monte Carlo approach is employed. In this respect, for each uncertain parameter, 1000 stochastic samples are generated randomly. To solve the stochastic problem, the following step-by-step procedure is considered:

- Step 1—Generation of 1000 samples for each uncertain optimization variable.
- Step 2—The stochastic generation of a set of solutions for optimization variables.
- Step 3—The selection of a set of optimization variables.
- Step 4—Test the grid's performance for 1000 samples using the Monte Carlo approach and checking all constraints for 1000 plausible scenarios.
- Step 5—The calculation of the objective function (if the constraints are met).
- Step 6—Check convergence and termination conditions. If it is converged, then go to step 9.
- Step 7—The change of optimization variables based on the optimization method.
- Step 8—Go to step 3.
- Step 9—End.

Table 4 outlines the results of the design of the microgrid with the incorporation of uncertainties.

Table 4. The results of optimal sizing of test grids incorporating uncertainties of demand and solar radiation and the consideration of the maximum cost of stochastic scenarios as the objective function.

Bus	Location	PV Size	Diesel Size	Number of Batteries	Cost (€)
33	2	395	215	42	70,342
69	6	400	214	39	69,214

In the case of the 33-bus test system, the cost of power generation is greatly increased. This increase is due to the conditions in which the load may be maximum and production may be minimum. Therefore, these conditions must also be taken into account in the problem. However, as can be seen, the cost has increased by nearly 49% compared to the case where uncertainty is not considered. Thus, it should be considered that in the presence of sources of intermittency, the cost of power supply from the network may increase remarkably, which conveys the importance of stochastic solution.

The increase in the system cost pertaining to the uncertainty will be 27%, which is nearly 20% less than the previous case. Therefore, the uncertainties are applied to the model using the average intermittencies in the system cost in all possible scenarios as an objective function. As can be seen, in the stochastic model, the cost of the system has increased by 26% compared to the deterministic case.

As the results of the 69-bus test system figure out, the cost of supply in this condition is significantly increased. This increase conveys a condition that the demand is at a high level and the solar radiation is at a low level. Therefore, the worst conditions must be contemplated in the design of the microgrid. However, as it is evident, the cost is drastically increased by 50% in comparison with the case of the deterministic model (regardless of uncertainties). Thus, it should be noted that the overall cost of the microgrid scheme may be enhanced extensively for uncertain problems.

It is important to pay attention to a critical question. The cost of supply has increased by 50% in order to include uncertainties. It conveys a high level of consumption and a low level of generation. The occurrence of such generation and consumption levels for this system at the targeted time is not a definitive and certain event. The question is whether it is sensible to increase the cost of the system by 50% for a forecast that has an occurrence probability of 10%.

In the procedure of seeking the optimum solution of uncertain parameters, for each suggested answer provided by the optimization program, all possible scenarios are taken into account, and the highest cost incurred to the system, among all scenarios, is regarded as the cost of the suggested answer. If the suggested answer does not satisfy even one or more constraints, that answer should be dismissed. Such a procedure profoundly mounts the system cost.

In order to tackle this problem, no answer should be dismissed, and the system cost must be obtained for all answers in as scenarios, even when they were unable to meet some constraints. Finally, the average cost of all scenarios is dedicated to the answer. Hence, if the suggested answer does not meet the constraints in some scenarios, the answer will not be dismissed. It is obvious that, for the problems with a few scenarios, the inclusion of scenarios, which used to be disregarded, does not have a remarkable impact on the solution of the problem. With consideration of the proposed condition, the optimal results of the simulation can be expressed in Table 5. This assumption has been led to an increase of 26% in the system cost for the incorporation of uncertainties in the model. It is just over 20% fewer than the normal approach. Thus, the uncertainties can be included in the design by incorporation of the average cost of the system. As can be seen, the system cost has increased by 26% in comparison with the deterministic approach.

Table 5. The results of optimal sizing of test grids incorporating uncertainties of demand and solar radiation and the consideration of the average cost of stochastic scenarios as the objective function.

Bus	Location	PV Size	Diesel Size	Number of Batteries	Cost (€)
33	2	395	253	37	59,802
69	2	400	265	34	59,325

5. Conclusions

In this study, a microgrid scheme including loads, PV panels, energy storage system, and the backup supply system of a diesel generator was designed. The object of the design was to minimize the cost of supply. In this microgrid, it was intended to not absorb any power from the distribution grid. However, if there is an excess of generation by intrinsic power resources of the microgrid, the excess power would be sold to the distribution network. The costs pertaining to PV panels, batteries, and diesel generator consists of purchasing and installation cost, operation and maintenance cost, and replacement cost. If the proposed scheme cannot supply the loads (generation deficiency) at some hours, the load-shedding measure must be imposed for the surplus demand. Just the same, the excess generation must be injected into the upstream network. In order to boost the performance of the design, the placement of the microgrid is so determined that the network power losses are minimized. Thus, the optimal design of the microgrid was performed at the same time as the placement of the microgrid. The problem solution was targeted to be stochastic. Hence, two upper and lower boundaries were dedicated to the amount of demand and solar irradiance. In order to solve the problem, the Monte Carlo approach with 6222 random samples is employed. The results of the 33-bus test system imply that the incorporation of uncertainties in the model has drastically increased the supply cost by about 49%. In order to avoid unnecessary risks, the averaging method is deployed, which boosts the risk-averse scheduling. In this case, the system cost has increased by 27% in comparison with the deterministic approach. The results of the 69-bus test system imply that the incorporation of uncertainties in the model has drastically increased the supply cost by about 50%. In order to avoid unnecessary risks, the averaging method is deployed, which boosts the risk-averse scheduling. In this case, the system cost has increased by 26% in comparison with the deterministic approach. The optimization is solved using the LAPO method, and the results are compared with the ABC algorithm. The results indicate the proper performance of the suggested algorithm in terms of convergence speed and accuracy.

Author Contributions: Conceptualization, A.F.N., H.S., H.N., B.V., Y.A. and M.B.; Methodology, A.F.N., H.S., H.N., B.V., Y.A. and M.B.; Supervision, A.F.N., H.N., B.V.; Validation, A.F.N., H.N., B.V.; Visualization, A.F.N., H.N., B.V.; Writing—original draft, A.F.N., H.S., H.N., B.V., Y.A. and M.B.; Writing—review & editing, A.F.N., H.S., H.N., B.V., Y.A. and M.B. All authors have read and agreed to the published version of the manuscript.

Funding: This research received no external funding.

Institutional Review Board Statement: Not applicable.

Informed Consent Statement: Not applicable.

Conflicts of Interest: The authors declare no conflict of interest.

References

1. Gil-González, W.; Garces, A.; Montoya, O.D.; Hernández, J.C. A Mixed-Integer Convex Model for the Optimal Placement and Sizing of Distributed Generators in Power Distribution Networks. *Appl. Sci.* **2021**, *11*, 627. [CrossRef]
2. Nafisi, H.; Agah, S.M.M.; Abyaneh, H.A.; Abedi, M. Two-Stage Optimization Method for Energy Loss Minimization in Microgrid Based on Smart Power Management Scheme of PHEVs. *IEEE Trans. Smart Grid* **2015**, *7*, 1268–1276. [CrossRef]
3. Nafisi, H.; Farahani, V.; Abyaneh, H.A.; Abedi, M. Optimal daily scheduling of reconfiguration based on mini-misation of the cost of energy losses and switching operations in microgrids. *IET Gener. Transm. Distrib.* **2015**, *9*, 513–522. [CrossRef]
4. Hosseini, S.J.A.D.; Moradian, M.; Shahinzadeh, H.; Ahmadi, S. Optimal Placement of Distributed Generators with Regard to Reliability Assessment using Virus Colony Search Algorithm. *Int. J. Renew. Energy Res.* **2018**, *8*, 714–723.
5. Hamzeh, M.; Vahidi, B.; Nematollahi, A.F. Optimizing Configuration of Cyber Network Considering Graph Theory Structure and Teaching-Learning-Based Optimization (GT-TLBO). *IEEE Trans. Ind. Inform.* **2018**, *15*, 2083–2090. [CrossRef]
6. Thatte, A.A.; Xie, L. Towards a Unified Operational Value Index of Energy Storage in Smart Grid Environment. *IEEE Trans. Smart Grid* **2012**, *3*, 1418–1426. [CrossRef]
7. Das, T.; Krishnan, V.; McCalley, J.D. Assessing the benefits and economics of bulk energy storage technologies in the power grid. *Appl. Energy* **2015**, *139*, 104–118. [CrossRef]
8. Rahiminejad, A.; Aranizadeh, A.; Vahidi, B. Simultaneous distributed generation and capacitor placement and sizing in radial distribution system considering reactive power market. *J. Renew. Sustain. Energy* **2014**, *6*, 43124. [CrossRef]
9. Moazzami, M.; Gharehpetian, G.B.; Shahinzadeh, H.; Hosseinian, S.H. Optimal locating and sizing of DG and D-STATCOM using Modified Shuffled Frog Leaping Algorithm. In Proceedings of the 2017 2nd Conference on Swarm Intelligence and Evolutionary Computation (CSIEC), Kerman, Iran, 7–9 March 2017; Institute of Electrical and Electronics Engineers (IEEE): Piscataway, NJ, USA, 2017; pp. 54–59.
10. Karunarathne, E.; Pasupuleti, J.; Ekanayake, J.; Almeida, D. Optimal Placement and Sizing of DGs in Distribution Networks Using MLPSO Algorithm. *Energies* **2020**, *13*, 6185. [CrossRef]
11. Cho, B.; Ryu, K.; Park, J.; Moon, W.; Cho, S.; Kim, J. A Selecting Method of Optimal Load on Time Varying Distribution System for Network Reconfiguration considering DG. *J. Int. Counc. Electr. Eng.* **2012**, *2*, 166–170. [CrossRef]
12. Aman, M.; Jasmon, G.; Bakar, A.; Mokhlis, H. A new approach for optimum simultaneous multi-DG distributed generation Units placement and sizing based on maximization of system loadability using HPSO (hybrid particle swarm optimization) algorithm. *Energy* **2014**, *66*, 202–215. [CrossRef]
13. Ameli, A.; Bahrami, S.; Khazaeli, F.; Haghifam, M.-R. A multiobjective particle swarm optimization for sizing and placement of DGs from DG owner's and distribution company's viewpoints. *IEEE Trans. Power Deliv.* **2014**, *29*, 1831–1840. [CrossRef]
14. Buaklee, W.; Hongesombut, K. Optimal DG Placement in a Smart Distribution Grid Considering Economic As-pects. *J. Electr. Eng. Technol.* **2014**, *9*, 1240–1247. [CrossRef]
15. Wang, Z.; Chen, B.; Wang, J.; Kim, J.; Begovic, M.M. Robust Optimization Based Optimal DG Placement in Microgrids. *IEEE Trans. Smart Grid* **2014**, *5*, 2173–2182. [CrossRef]

16. Alotaibi, M.; Almutairi, A.; Salama, M. Effect of wind turbine parameters on optimal DG placement in power dis-tribution systems. In Proceedings of the 2016 IEEE Electrical Power and Energy Conference (EPEC), Ottawa, ON, Canada, 12–14 October 2016; Institute of Electrical and Electronics Engineers (IEEE): Piscataway, NJ, USA, 2016.
17. Hengsritawat, V.; Tayjasanant, T.; Nimpitiwan, N. Optimal sizing of photovoltaic distributed generators in a dis-tribution system with consideration of solar radiation and harmonic distortion. *Int. J. Electr. Power Energy Syst.* **2012**, *39*, 36–47. [CrossRef]
18. Carpentiero, V.; Langella, R.; Testa, A. Hybrid wind-diesel stand-alone system sizing accounting for component expected life and fuel price uncertainty. *Electr. Power Syst. Res.* **2012**, *88*, 69–77. [CrossRef]
19. Hakimi, S.M.; Moghaddas-Tafreshi, S.M. Optimal sizing of a stand-alone hybrid power system via particle swarm optimization for Kahnouj area in south-east of Iran. *Renew. Energy* **2009**, *34*, 1855–1862. [CrossRef]
20. Ma, T.; Yang, H.; Lu, L. A feasibility study of a stand-alone hybrid solar-wind-battery system for a remote island. *Appl. Energy* **2014**, *121*, 149–158. [CrossRef]
21. Hassan, A.S.; Sun, Y.; Wang, Z. Multi-objective for optimal placement and sizing DG units in reducing loss of power and enhancing voltage profile using BPSO-SLFA. *Energy Rep.* **2020**, *6*, 1581–1589. [CrossRef]
22. Reddy, P.D.P.; Reddy, V.V.; Manohar, T.G. Ant Lion optimization algorithm for optimal sizing of renewable energy resources for loss reduction in distribution systems. *J. Electr. Syst. Inf. Technol.* **2018**, *5*, 663–680. [CrossRef]
23. Alzahrani, A.; Alharthi, H.; Khalid, M. Minimization of power losses through optimal battery placement in a dis-tributed network with high penetration of photovoltaics. *Energies* **2020**, *13*, 140.
24. Farh, H.M.H.; Al-Shaalan, A.M.; Eltamaly, A.M.; Al-Shamma'A, A.A. A Novel Crow Search Algorithm Auto-Drive PSO for Optimal Allocation and Sizing of Renewable Distributed Generation. *IEEE Access* **2020**, *8*, 27807–27820. [CrossRef]
25. HassanzadehFard, H.; Jalilian, A. Optimal sizing and location of renewable energy based DG units in distribution systems considering load growth. *Int. J. Electr. Power Energy Syst.* **2018**, *101*, 356–370. [CrossRef]
26. Hemeida, M.G.; Ibrahim, A.A.; Mohamed, A.-A.A.; Alkhalaf, S.; El-Dine, A.M.B. Optimal allocation of distributed generators DG based Manta Ray Foraging Optimization algorithm (MRFO). *Ain Shams Eng. J.* **2021**, *12*, 609–619. [CrossRef]
27. Jalili, A.; Taheri, B. Optimal Sizing and Sitting of Distributed Generations in Power Distribution Networks Using Firefly Algorithm. *Technol. Econ. Smart Grids Sustain. Energy* **2020**, *5*, 1–14. [CrossRef]
28. Lata, P.; Vadhera, S. Optimal placement and sizing of energy storage systems to improve the reliability of hybrid power distribution network with renewable energy sources. *J. Stat. Manag. Syst.* **2020**, *23*, 17–31. [CrossRef]
29. Mukhopadhyay, B.; Das, D. Multi-objective dynamic and static reconfiguration with optimized allocation of PV-DG and battery energy storage system. *Renew. Sustain. Energy Rev.* **2020**, *124*, 109777. [CrossRef]
30. Shahinzadeh, H.; Moradi, J.; Gharehpetian, G.B.; Nafisi, H.; Abedi, M. IoT Architecture for Smart Grids. In Proceedings of the 2019 International Conference on Protection and Automation of Power System (IPAPS), Tehran, Iran, 8–9 January 2019; Institute of Electrical and Electronics Engineers (IEEE): Piscataway, NJ, USA, 2019; pp. 22–30.
31. Shahinzadeh, H.; Moradi, J.; Gharehpetian, G.B.; Nafisi, H.; Abedi, M. Internet of Energy (IoE) in Smart Power Systems. In Proceedings of the 2019 5th Conference on Knowledge Based Engineering and Innovation (KBEI), Tehran, Iran, 28–29 February 2019; Institute of Electrical and Electronics Engineers (IEEE): Piscataway, NJ, USA, 2019; pp. 627–636.
32. Ebeed, M.; Ali, A.; Mosaad, M.I.; Kamel, S. An Improved Lightning Attachment Procedure Optimizer for Optimal Reactive Power Dispatch With Uncertainty in Renewable Energy Resources. *IEEE Access* **2020**, *8*, 168721–168731. [CrossRef]
33. Youssef, H.; Kamel, S.; Ebeed, M. Optimal Power Flow Considering Loading Margin Stability Using Lightning Attachment Optimization Technique. In Proceedings of the 2018 Twentieth International Middle East Power Systems Conference (MEPCON, Nasr City, Egypt, 18–20 December 2018; Institute of Electrical and Electronics Engineers (IEEE): Piscataway, NJ, USA, 2018; pp. 1053–1058.
34. Ramadan, A.; Ebeed, M.; Kamel, S.; Nasrat, L. Optimal Allocation of Renewable Energy Resources Considering Uncertainty in Load Demand and Generation. In Proceedings of the 2019 IEEE Conference on Power Electronics and Renewable Energy (CPERE), Aswan, Egypt, 23–25 October 2019; Institute of Electrical and Electronics Engineers (IEEE): Piscataway, NJ, USA, 2019; pp. 124–128.
35. Ebeed, M.; Alhejji, A.; Kamel, S.; Jurado, F. Solving the Optimal Reactive Power Dispatch Using Marine Predators Algorithm Considering the Uncertainties in Load and Wind-Solar Generation Systems. *Energies* **2020**, *13*, 4316. [CrossRef]
36. Ramadan, A.; Ebeed, M.; Kamel, S.; Nasrat, L. Optimal power flow for distribution systems with uncertainty. In *Uncertainties in Modern Power Systems*; Elsevier BV: Amsterdam, The Netherlands, 2021; pp. 145–162.
37. Ebeed, M.; Aleem, S.H.E.A. Overview of uncertainties in modern power systems: Uncertainty models and methods. In *Uncertainties in Modern Power Systems*; Elsevier BV: Amsterdam, The Netherlands, 2021; pp. 1–34.
38. Yang, Y.; Li, S.; Li, W.; Qu, M. Power load probability density forecasting using Gaussian process quantile regres-sion. *Appl. Energy* **2018**, *213*, 499–509. [CrossRef]
39. Sirat, A.P.; Mehdipourpicha, H.; Zendehdel, N.; Mozafari, H. Sizing and Allocation of Distributed Energy Resources for Loss Reduction using Heuristic Algorithms. In Proceedings of the 2020 IEEE Power and Energy Conference at Illinois (PECI), Champaign, IL, USA, 20–21 February 2020; Institute of Electrical and Electronics Engineers (IEEE): Piscataway, NJ, USA, 2020; pp. 1–6.
40. Nematollahi, A.F.; Rahiminejad, A.; Vahidi, B. A novel physical based meta-heuristic optimization method known as Lightning Attachment Procedure Optimization. *Appl. Soft Comput.* **2017**, *59*, 596–621. [CrossRef]

41. Nematollahi, A.F.; Rahiminejad, A.; Vahidi, B.; Askarian, H.; Safaei, A. A new evolutionary-analytical two-step optimization method for optimal wind turbine allocation considering maximum capacity. *J. Renew. Sustain. Energy* **2018**, *10*, 043312. [CrossRef]
42. Nematollahi, A.F.; Rahiminejad, A.; Vahidi, B. A novel multi-objective optimization algorithm based on Lightning Attachment Procedure Optimization algorithm. *Appl. Soft Comput.* **2019**, *75*, 404–427. [CrossRef]
43. Nematollahi, A.F.; Dadkhah, A.; Gashteroodkhani, O.A.; Vahidi, B. Optimal sizing and siting of DGs for loss reduction using an iterative-analytical method. *J. Renew. Sustain. Energy* **2016**, *8*, 055301. [CrossRef]
44. El-Zonkoly, A. Optimal placement of multi-distributed generation units including different load models using particle swarm optimization. *Swarm Evol. Comput.* **2011**, *1*, 50–59. [CrossRef]
45. El-Zonkoly, A. Optimal placement and schedule of multiple grid connected hybrid energy systems. *Int. J. Electr. Power Energy Syst.* **2014**, *61*, 239–247. [CrossRef]

applied sciences

MDPI

Article

Development and Implementation of a Novel Optimization Algorithm for Reliable and Economic Grid-Independent Hybrid Power System

Mohammed Kharrich [1], Omar Hazem Mohammed [2], Salah Kamel [3], Ali Selim [3,4], Hamdy M. Sultan [5], Mohammed Akherraz [1] and Francisco Jurado [4,*]

1 Department of Electrical Engineering, Mohammadia School of Engineers, Mohammed V University, Ibn Sina Street P.B 765, Rabat 10090, Morocco; mohammedkharrich@research.emi.ac.ma (M.K.); akherraz@emi.ac.ma (M.A.)
2 Department of Technical Power Engineering, Technical College, Northern Technical University, Mosul 41002, Iraq; omar.hazem@ntu.edu.iq
3 Department of Electrical Engineering, Faculty of Engineering, Aswan University, Aswan 81542, Egypt; skamel@aswu.edu.eg (S.K.); ali.selim@aswu.edu.eg (A.S.)
4 Department of Electrical Engineering, University of Jaén, 23700 EPS Linares Jaén, Spain
5 Electrical Engineering Department, Faculty of Engineering, Minia University, Minia 61111, Egypt; hamdy.soltan@mu.edu.eg
* Correspondence: fjurado@ujaen.es

Received: 17 August 2020; Accepted: 16 September 2020; Published: 21 September 2020

Abstract: Recently, fast uptake of renewable energy sources (RES) in the world has introduced new difficulties and challenges; one of the most important challenges is providing economic energy with high efficiency and good quality. To reach this goal, many traditional and smart algorithms have been proposed and demonstrated their feasibility in obtaining the optimal solution. Therefore, this paper introduces an improved version of Bonobo Optimizer (BO) based on a quasi-oppositional method to solve the problem of designing a hybrid microgrid system including RES (photovoltaic (PV) panels, wind turbines (WT), and batteries) with diesel generators. A comparison between traditional BO, the Quasi-Oppositional BO (QOBO), and other optimization techniques called Harris Hawks Optimization (HHO), Artificial Electric Field Algorithm (AEFA) and Invasive Weed Optimization (IWO) is carried out to check the efficiency of the proposed QOBO. The QOBO is applied to a stand-alone hybrid microgrid system located in Aswan, Egypt. The results show the effectiveness of the QOBO algorithm to solve the optimal economic design problem for hybrid microgrid power systems.

Keywords: economic energy; Bonobo Optimizer; hybrid renewable energy system; microgrid; PV panels; wind turbine; energy storage

1. Introduction

Despite the steady increase in electric power production, it is still below the required level, due to the increase in load demand caused by the population increase as well as the increased use of technology in the residential, industrial and agricultural fields. According to the International Energy Agency (IEA), the global electricity demand will grow at an annual rate of 2.1% until 2040. This increases electricity's share in the total energy consumption to 24% in 2040. It is expected that renewable energy sources (RES) will face a significant increase in global investment in the coming years, to cover more than half of the energy consumption in the world by 2040. These energies will make up for the shortfall in electrical energy production and contribute to a reduction in carbon dioxide emissions in the atmosphere, thereby reducing pollution significantly [1–3].

In order to invest in RES to optimize electrical energy production and raise the efficiency of the systems, many studies in the world recommend combining different technologies to form hybrid renewable energy systems (HRES) [4,5]. Consequently, these sources complement each other, support the national grid, and reduce the use of traditional power plants depending on fossil fuels that release greenhouse gases and pollute the environment [6]. However, the design of these hybrid systems needs sophisticated programs and smart algorithms capable of reaching the optimal solution taking into consideration all the conditions and constraints such as reliability aspects, economic cost, sensitivity factors, availability of RES, etc. [1,2,7,8].

Several studies have been conducted on the technical and economic feasibility of hybrid systems in past years to determine their viability. Many of these studies have used different modeling of HRES, and they have applied different algorithms and various software tools to achieve their goals. According to the literature, these challenges still exist and are the focus of a lot of research, especially on finding the best algorithms and modern techniques in reaching the optimal solutions of the optimization problem of finding the optimal sizing of the installed capacities of the components of HRES [9–15].

In [16], the pre-feasibility analysis of a stand-alone energy system using HRES including renewable and conventional energy sources was applied using HOMER software in Newfoundland, Canada. In one of the earlier studies [17], the authors conducted a feasibility study of generating electricity using RES for a hybrid system in a stand-alone village in Chhattisgarh, India. In [18], the authors introduce a realistic solution for energy demand from a hybrid power system consists of wind turbines (WT), photovoltaics (PV), and battery energy storage systems (BESS). Through a real measurement of meteorological data in 2017, concerning especially the wind speed, solar radiation and temperature, the output power of the proposed hybrid system is calculated. Load satisfaction is considered to evaluate the feasibility of the system. The optimum solution is found using the linear TORSCHE optimization technique, while a comparative study between PV/WT/battery and PV/WT has been accomplished and an economic analysis was presented. As a result, the hybrid PV/WT/battery is proved more economical than using each system individually.

Xiao Xu et al. [19] designed and investigated a hybrid PV/WT/hydropower/pump storage as a case of study. The optimal configuration of the HRES is found using a techno-economic index that respects the maximum Loss of Power Supply Probability (LPSP) and minimum investment cost. The Multi-Objective Particle Swarm Optimization (MOPSO) is used to trade off analysis between two objectives. Besides, the curtailment rate (CR) of the WT and PV are taken into consideration due to policy requirements. The authors in [20] proposed an optimized design of an energy system featuring the highest penetration of renewable energy. This system is composed of WT, PV, geothermal, diesel, and BESS; otherwise, the system is obtained respecting the technological and financial feasibility constraints. The model developed is based on weather and electric demand data measured to reach the optimal sizing of the hybrid system. Three objective functions are conflicting, which are the Net Present Cost (NPC), renewable energy fraction and the energy index of reliability.

In [21], the authors implemented and compared three algorithms to find the optimal design of a hybrid WT/PV/Biomass/BESS energy system. Based on the obtained results, the Harmony Search Algorithm (HSA) was faster and efficient in the convergence, compared to Jaya and PSO optimization algorithms. The techno-economic study has been implemented to have the optimal unit sizing of the HRES, which guaranteed a cost-effective, efficient, and reliable power supply for the customers of electric energy. The constraints are chosen to enhance the reliability and efficiency of the hybrid system, using the LPSP and the energy fraction factors.

In this paper, a new smart algorithm named Bonobo Optimizer [22] was employed and improved using a quasi-oppositional method, and the modified Quasi Oppositional BO (QOBO) was utilized for optimal economic designing of a stand-alone microgrid hybrid system in Aswan, Egypt, where the hybrid system consists of RES (PV panels, WT and BESS) with diesel generators. Then, the results were compared between the traditional and improved BO. This proved the ability of the QOBO algorithm to reach the optimal solution in a shorter time and with better efficiency compared to the traditional BO algorithm. Other algorithms, namely Harris Hawks Optimization, Artificial Electric Field Algorithm and Invasive Weed Optimization are applied, and the results are compared where the efficiency of the QOBO algorithm has been proved. Additionally, a sensitivity analysis of the proposed systems scenarios was performed to obtain the optimal solution.

2. Mathematical Description of the Proposed Hybrid System Components

The schematic diagram of the suggested HRES is shown in Figure 1. Four scenarios are applied, which include the PV power plant, WT power plant, diesel generator, Biomass, BESS and inverter.

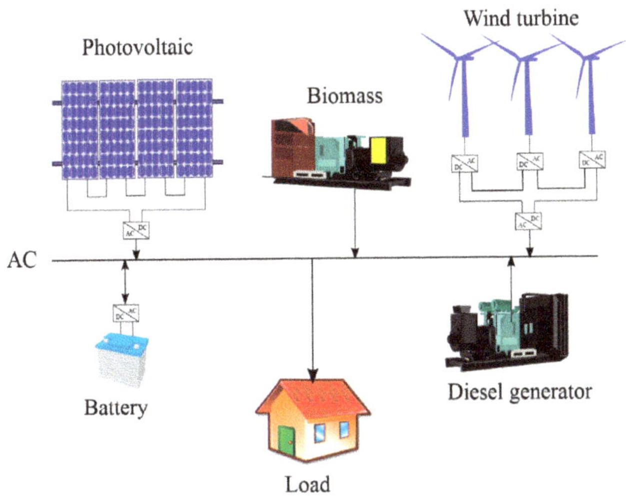

Figure 1. Configuration of the proposed microgrid hybrid energy system.

Two strategies are adopted in this paper; the first is the biomass/PV as shown in Figure 2 and the second uses the PV or WT or both as in Figure 3. The main strategy steps for the operation of the proposed system can be explained as follows:

- The PV and WT are used first as a principal power source and served the load needs.
- The battery is used when the PV and WT cannot serve it.
- The diesel system is working when the battery storage system is empty and starts when the need is higher than 30% of its nominal power.

2.1. PV System

The PV system is considered as a number of cells connected in series. The output power of the PV system is presented based on many parameters as introduced in Equation (1) [23]:

$$P_{pv} = I(t) \times \eta_{pv}(t) \times A_{pv} \tag{1}$$

where I represents the solar irradiation, A_{pv} represents the area covered with PV modules and η_{pv} is the efficiency of the PV system that can be calculated as follows:

$$\eta_{pv}(t) = \eta_r \times \eta_t \times \left[1 - \beta \times (T_a(t) - T_r) - \beta \times I(t) \times \left(\frac{NOCT - 20}{800}\right) \times (1 - \eta_r \times \eta_t)\right] \tag{2}$$

where $NOCT$ is the nominal operating cell temperature (°C), η_r is the reference efficiency, η_t is the efficiency of the maximum power point tracking (MPPT) equipment, β is the temperature coefficient, T_a is the ambient temperature (°C), T_r is the solar cell reference temperature (°C).

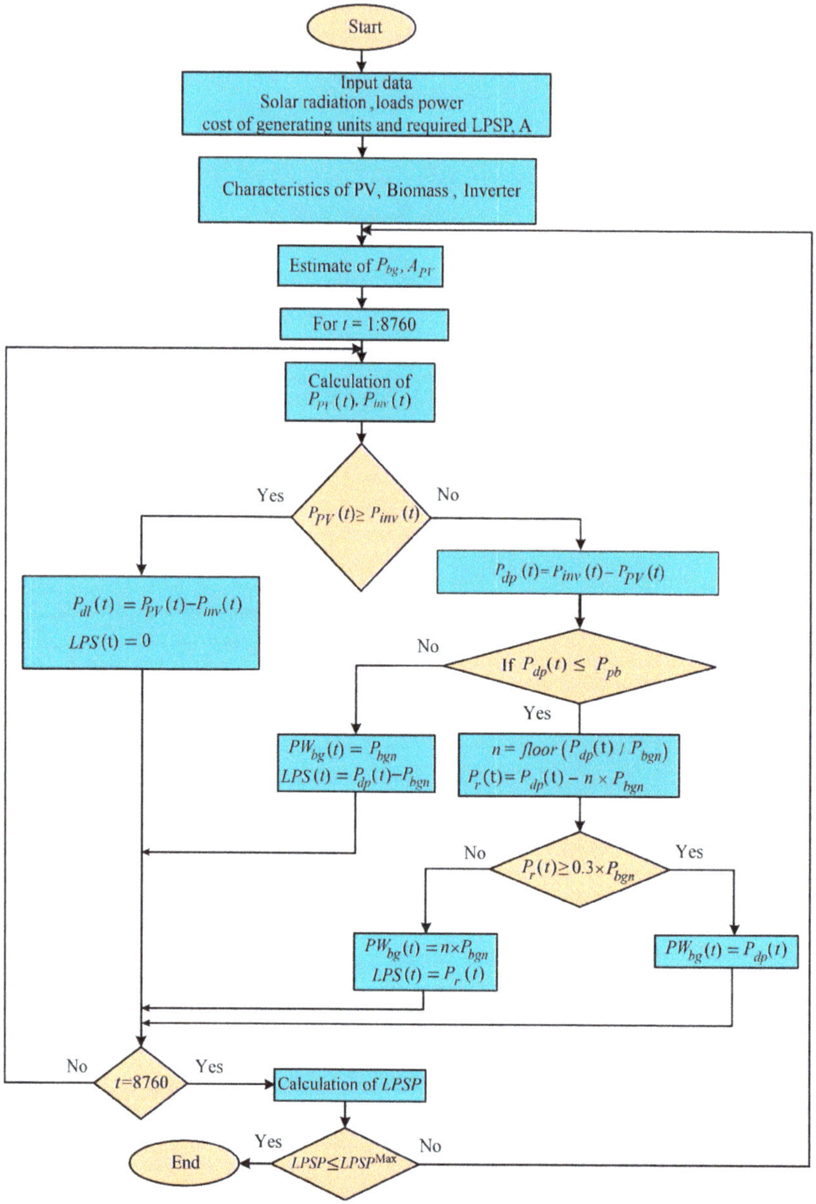

Figure 2. Power management of the PV/Biomass hybrid renewable energy sources (RES).

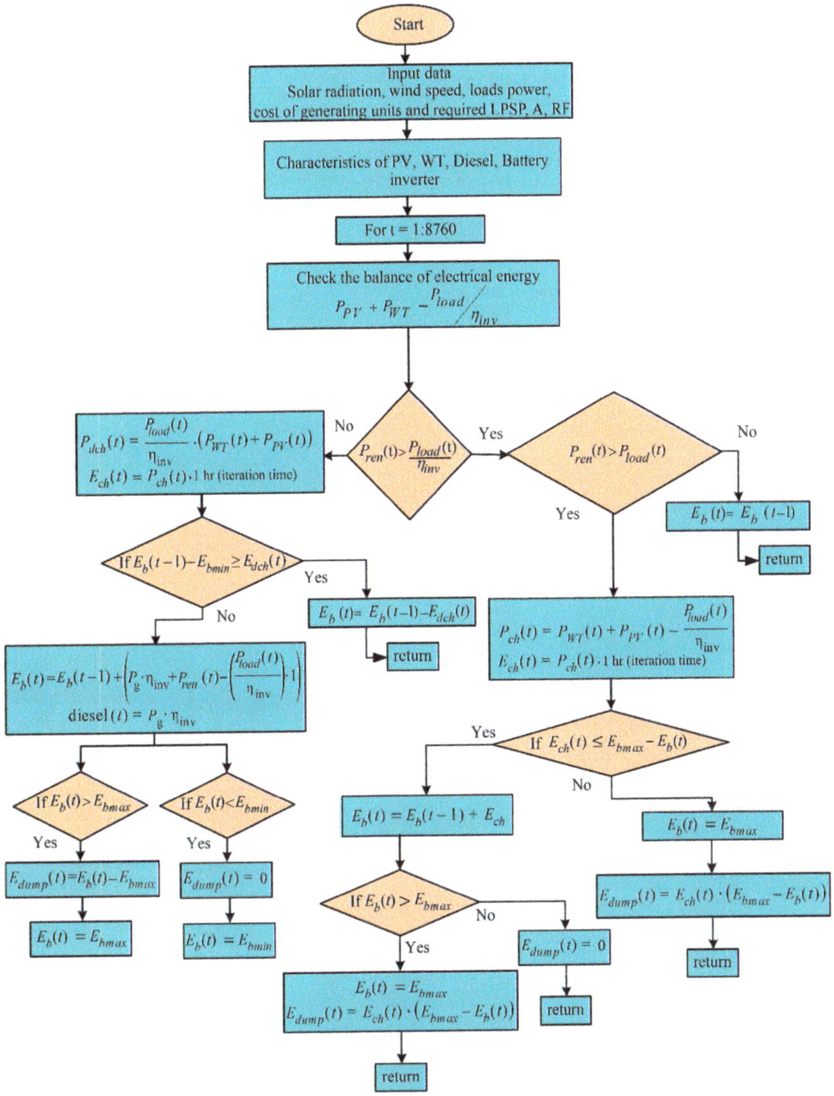

Figure 3. Power management of the PV/WT/diesel/battery energy storage system (BESS), PV/diesel/BESS and WT/diesel/BESS hybrid RES.

2.2. Wind Energy System

Based on the basics of aerodynamics, wind power can be presented as [24]:

$$P_{wind} = \begin{cases} 0, & V(t) \leq V_{ci},\ V(t) \geq V_{co} \\ a \times V(t)^3 - b \times P_r, & V_{ci} < V(t) < V_r \\ P_r, & V_r \leq V(t) < V_{co} \end{cases} \tag{3}$$

where V represents wind speed, P_r is the rated power of wind, V_{ci}, V_{co} and V_r are the cut-in, cut-out, and rated wind speed, respectively. a and b are two constants, which can be expressed as:

$$\begin{cases} a = P_r/\left({V_r}^3 - {V_{ci}}^3\right) \\ b = {V_{ci}}^3/\left({V_r}^3 - {V_{ci}}^3\right) \end{cases} \tag{4}$$

The rated power of wind is calculated as given in the following equation:

$$P_r = \frac{1}{2} \times \rho \times A_{wind} \times C_p \times {V_r}^3 \tag{5}$$

where ρ represents the air density, A_{wind} is the swept area of the wind turbine, C_p is the maximum power coefficient ranging from 0.25% to 0.45%.

2.3. Biomass System

The biomass system is a renewable energy system, which produces power as given in Equation (6) [23].

$$P_{BM} = \frac{Total_{bio} \times 1000 \times CV_{bio} \times \eta_{bio}}{8760 \times O_{time}} \tag{6}$$

where $Total_{bio}$ is the total organic material of biomass, CV_{bio} is the calorific value of the organic material (20 MJ/kg), η_{bio} is the biomass efficiency, which is taken as 24% and O_{time} presents the operating hours each day.

2.4. Diesel System

The diesel generator is used as a back-up, working just in case there is a need, is connected directly with the load, and starts when the battery is fully discharged and the load is more than 30% of its rated capacity. The model of the diesel generator regarding its output power is presented by the following Equation [25]:

$$P_{dg} = \frac{F_{dg}(t) - A_g \times P_{dg,out}}{B_g} \tag{7}$$

where F_{dg} is fuel consumption, $P_{dg,out}$ is the output power of diesel generator, A_g and B_g are the constants of the linear consumption of the fuel.

2.5. BESS System

The battery energy storage system (BESS) is a mandatory element for the isolated hybrid systems. BESS is charged in the periods of power excess and discharged when the load increases. The capacity of the BESS is expressed as follows [25]:

$$C_{bat} = \frac{E_l \times AD}{DOD \times \eta_{inv} \times \eta_b} \tag{8}$$

where E_l is the load demand, AD represents the autonomy daily of the battery, DOD is the depth of discharge of the battery system, η_{inv} and η_b are the battery and inverter efficiency, respectively.

3. Formulation of the Optimization Problem

3.1. Net Present Cost

The objective function in the optimization model is the minimization for the Net Present Cost (NPC) which is the pillar factor considered for any project design; it is counted as a sum of all components costs including the capital (C), operation and maintenance (OM) and replacement costs (R), considering also the fuel cost of the diesel $\left(FC_{dg}\right)$, taking into account the interest rate (i_r), inflation rate (δ), and escalation rate (μ) and the predefined project lifetime (N). The NPC modeling is expressed as follows [23,24]:

$$NPC = C + OM + R + FC_{dg} \tag{9}$$

3.1.1. PV and WT Costs

The costs of PV and WT are presented in a similar concept, their capital cost is expressed based on its initial cost ($\lambda_{PV,WT}$) and its area ($A_{PV,WT}$), the capital cost is as follows [26]:

$$C_{PV,WT} = \lambda_{PV,WT} \times A_{PV,WT} \tag{10}$$

The operation and maintenance costs are expressed as [26]:

$$OM_{PV,WT} = \theta_{PV,WT} \times A_{PV,WT} \times \sum_{i=1}^{N} \left(\frac{1+\mu}{1+i_r}\right)^i \tag{11}$$

where $\theta_{PV,WT}$ is the annual operation and maintenance costs for any components. The replacement costs are considered null because the project lifetime and the *PV* or *WT* lifetime are the same.

3.1.2. Diesel Generator Costs

The costs of the diesel generator are presented as follows [27]:

$$C_{dg} = \lambda_{dg} \times P_{dg} \tag{12}$$

$$OM_{dg} = \theta_{dg} \times N_{run} \times \sum_{i=1}^{N} \left(\frac{1+\mu}{1+i_r}\right)^i \tag{13}$$

$$R_{diesel} = R_{dg} \times P_{dg} \times \sum_{i=7,14\ldots} \left(\frac{1+\delta}{1+i_r}\right)^i \tag{14}$$

$$C_f(t) = p_f \times F_{dg}(t) \tag{15}$$

$$FC_{dg} = \sum_{t=1}^{8760} C_f(t) \times \sum_{i=1}^{N} \left(\frac{1+\delta}{1+i_r}\right)^i \tag{16}$$

where C_{dg} is the capital cost, λ_{dg} is the initial cost of the diesel generator for each KW, OM_{dg} represent the actual O&M cost, θ_{dg} is the annual O&M cost of diesel, N_{run} is the number of operating hours of diesel generator per year, R_{diesel} is the diesel generator replacement cost, R_{dg} represents the annual replacement cost of diesel generator, p_f is the fuel cost, F_{dg} is the annual consumption of fuel and FC_{dg} is the total fuel cost.

3.1.3. BESS Costs

The capital and O&M (containing the replacement) costs of the BESS are expressed as follows [26]:

$$C_{BESS} = \lambda_{bat} \times C_{bat} \tag{17}$$

$$OM_{BESS} = \theta_{bat} \times C_{bat} \times \sum_{i=1}^{T_B} \left(\frac{1+\mu}{1+\delta}\right)^{(i\_1)N_{bat}} \tag{18}$$

where λ_{bat} is the BESS initial cost and θ_{bat} is the annual O&M cost of BESS.

3.1.4. Biomass Costs

The biomass costs are presented as follows [28]:

$$C_{bg} = \lambda_{bg} \times P_{bg} \tag{19}$$

$$OM_{bg} = \theta_1 \times P_{bg} \times \sum_{i=1}^{N} \left(\frac{1+\mu}{1+i_r}\right)^i + \theta_2 \times P_w \times \sum_{i=1}^{N} \left(\frac{1+\mu}{1+i_r}\right)^i \quad (20)$$

where λ_{bg} is the biomass initial cost, θ_1 is the annual fixed O&M cost and θ_2 is the variable O&M cost of the biomass system, and P_w is the annual energy generated by the Biomass system (kWh/Year).

3.1.5. Inverter Costs

The inverter capital and O&M costs are presented as follows [27]:

$$C_{inv} = \lambda_{inv} \times P_{inv} \quad (21)$$

$$OM_{Inv} = \theta_{Inv} \times \sum_{i=1}^{N} \left(\frac{1+\mu}{1+i_r}\right)^i \quad (22)$$

where λ_{inv} is the inverter initial cost and θ_{Inv} is the annual O&M cost of the inverter.

3.2. Levelized Cost of Energy

The Levelized Cost of Energy (*LCOE*) is a critical factor. The consumers do not care about project cost or its lifetime, but their interest is to know how much to pay for each kilowatt-hour of consumption. Therefore, the *LCOE* is a measure of the average *NPC* over its lifetime, its equation is expressed as follows [25]:

$$LCOE = \frac{NPC \times CRF}{\sum_{t=1}^{8760} P_{load}(t)} \quad (23)$$

where P_{load} is the load demand; *CRF* is the capital recovery factor used to convert the initial cost to an annual capital cost, and is expressed as follow:

$$CRF(ir, R) = \frac{i_r \times (1+i_r)^R}{(1+i_r)^R - 1} \quad (24)$$

where *R* denotes the lifetime of the hybrid system.

3.3. Loss of Power Supply Probability

The loss of power supply probability (LPSP) is a technical factor used to express the reliability of the system. The LPSP is expressed as follows [25]:

$$LPSP = \frac{\sum_{t=1}^{8760} \left(P_{load}(t) - P_{pv}(t) - P_{wind}(t) + P_{dg,out}(t) + P_{bmin}\right)}{\sum_{t=1}^{8760} P_{load}(t)} \quad (25)$$

3.4. Renewable Energy Fraction

The transfer from classical electricity production to renewable energy projects was not easy. The majority introduced RES partially, while the objective is to use all projects with 100% renewable energy. Therefore, the renewable energy factor is dedicated to calculating the percentage of the renewable energy used. The renewable energy fraction (*RF*) is expressed as follows [25]:

$$RF = \left(1 - \frac{\sum_{t=1}^{8760} P_{dg,out}(t)}{\sum_{t=1}^{8760} P_{re}(t)}\right) \times 100 \quad (26)$$

where P_{re} represents the total power from RES.

3.5. Availability Index

The availability index (A) is calculated to predict customer satisfaction. The availability index measures the energy converted to the load while confirming the ability of the designing system of the project. The availability index is calculated as follows [23]:

$$A = 1 - \frac{DMN}{\sum_{t=1}^{8760} P_{load}(t)} \tag{27}$$

$$DMN = P_{bmin}(t) - P_b(t) - \left(P_{pv}(t) + P_{wind}(t) + P_{dg,out}(t) - P_{load}(t)\right) \times u(t) \tag{28}$$

while, u will be equal to 1 when the load is not satisfied, and 0 when the load is satisfied.

3.6. Constraints

The constraints are presented to achieve the desired system design. In this microgrid system, the constraints are shown as follows:

$$\begin{gathered} 0 \le A_{pv} \le A_{pv}^{max}, \\ 0 \le A_{wind} \le A_{wind}^{max}, \\ 0 \le P_{dgn} \le P_{dgn}^{max}, \\ 0 \le P_{Cap\_bat} \le P_{Cap\_bat}^{max} \\ LPSP \le LPSP^{max}, \\ RF^{min} \le RF, \\ A^{min} \le A \\ AD^{min} \le AD \end{gathered} \tag{29}$$

4. Algorithms

In this section, the conventional BO and proposed QOBO are illustrated. In addition, both algorithms are compared with well-known optimization techniques (HHO, AEFA and IWO) which are briefly described in Appendix A.

4.1. Bonobo Optimizer

Bonobo optimizer is a new optimization algorithm that was proposed in [22]. In BO, the social reproductive behavior of the bonobo is modeled based on four mating strategies: promiscuous, restrictive, consortship, and extra-group mating. These mating strategies are subjected to the living condition of the bonobo, hence two terms named positive phase (PP) and negative phase (NP) have been used to present the situations of this life. In this framework, PP describes the peaceful living in which the mating can be done. On the contrary, NP expresses a hard life. In the BO, each solution is called X_B and the best solution is X_B^{α}. The mathematical modeling of the BO algorithm is presented in the following subsections.

4.1.1. Bonobo Selection Using Fission–Fusion Strategy

The solutions update of the BO algorithm depends on the mating strategies subjected to the current phase. However, a bonobo should be selected before each mating based on the fission–fusion social group strategy. As noted, the bonobo community lives in small groups with different sizes (random and unpredictable) for a few days and the communities rejoined again to the main community. Hence, based on this behavior, a bonobo for mating can be selected. The mathematical formulation for the maximum number of these temporary subgroups N_{sub} can be expressed as follows:

$$N_{sub} = max(2, (\varepsilon_{sub} \times N)) \tag{30}$$

where N is the total number of the population and ε_{sub} denotes the sub-group size factor. To find the selected bonobo X_B^P for mating with X_B^i to create a new bonobo X_B^{new}, if the best bonobo in the subgroup in terms of the fitness function is better than the X_B^i, then it is selected as X_B^P, else a random one should be selected form the subgroup.

4.1.2. Creation of New Bonobo

After achieving the selected bonobo X_B^P, four mating strategies are used in the BO algorithm to create a new bonobo X_B^{new} based on the current phase (PP or NP). For the PP case, promiscuous and restrictive mating have higher probabilities (ρ_{ph}) for occurrence. On the contrary, in NP, the probabilities (ρ_{ph}) of consortship mating and extra-group mating are higher.

Promiscuous and Restrictive Mating

In this mating strategy, the new bonobo can be created using the following equation:

$$X_B^{new} = X_B^i + r_1 \times S_{coef}^{\alpha} \times \left(X_B^{\alpha} - X_B^i\right) + (1 - r_1) \times S_{coef}^{P} \times C_{flag} \times \left(X_B^i - X_B^P\right) \quad (31)$$

where r_1 is a random number between [0, 1]. S_{coef}^{α} and S_{coef}^{P} are the sharing coefficients for the alpha bonobo X_B^{α} and the selected bonobo X_B^P, respectively. C_{flag} is a flag value that equals −1 or 1 for restrictive and promiscuous mating, respectively. A controlling parameter in terms of the phase probability ρ_{ph} is used to adopt the mating strategy. Initially, ρ_{ph} is set to 0.5. Hence, if a random number r is found to be less than or equal to ρ_{ph}, a new bonobo is created based on promiscuous and restrictive mating, otherwise, consortship mating and extra-group mating can be used.

Consortship and Extra-Group Mating

If r is greater than ρ_{ph}, consortship and extra-group mating can occur. However, a new random number r_2 between [0, 1] is used with a probability of extra-group mating ρ_{xg} to represent the occurrence of extra-group mating when r_2 is less than or equal to ρ_{xg} as follows [22,29]:

$$X_B^{new} = \begin{cases} X_B^i + \beta_1 \times \left(X_{max}^i - X_B^i\right), & X_B^{\alpha} \geq X_B^i, and\ r_4 \leq \rho_d \\ X_B^i - \beta_2 \times \left(X_B^i - X_{min}^i\right), & X_B^{\alpha} \geq X_B^i, and\ r_4 > \rho_d \\ X_B^i - \beta_1 \times \left(X_B^i - X_{min}^i\right), & X_B^{\alpha} < X_B^i, and\ r_4 \leq \rho_d \\ X_B^i + \beta_2 \times \left(X_{max}^i - X_B^i\right), & X_B^{\alpha} < X_B^i, and\ r_4 > \rho_d \end{cases} \quad (32)$$

$$\beta_1 = e^{\left(r_4^2 + r_4 - \frac{2}{r_4}\right)} \qquad \beta_2 = e^{\left(-r_4^2 + 2r_4 - \frac{2}{r_4}\right)} \quad (33)$$

where r_3 and r_4 are random numbers between [0, 1] and $r_4 \neq 0$. ρ_d is a directional probability with initial value which equals 0.5. β_1 *and* β_2 are intermediate parameters between [0, 1]. X_{min}^i and X_{min}^i are the values of the upper and lower boundary.

If r_2 is greater than ρ_{xg}, a new bonobo can be created using the consortship mating strategy as follows:

$$X_B^{new} = \begin{cases} X_B^i + C_{flag} \times e^{-r_5} \times \left(X_B^i - X_B^P\right), & C_{flag} = 1\ or\ r_6 \leq \rho_d \\ X_B^P, & Otherwise \end{cases} \quad (34)$$

where r_5 and r_6 are two random numbers.

4.1.3. Parameter Updating

The BO's parameters are updated during the iterative process based on the best solution X_B^{α} at each iteration, where if there is an improvement in the final solution compared to the previous iteration, the BO's parameters can be updated in the following way.

The negative phase count is set to zero ($NP_{cont} = 0$) and the positive phase count grows by increments of one ($PP_{cont} = PP_{cont} + 1$). In addition, $\rho_{xg} = \rho_{xg\_initial}$ and $\rho_{ph} = 0.5 + Cp$ where Cp is the amount of the change in the phase, and can be calculated as $Cp = \min(0.5,\ PP_{cont} \times rcp)$ where rcp is the rate of the change in the phase. Moreover $\rho_d = \rho_{ph}$ and

$$\varepsilon_{sub} = min\left(\varepsilon_{sub\_max},\ \left(\varepsilon_{sub_{initial}} + PP_{cont} \times rcp^2\right)\right) \tag{35}$$

where $\varepsilon_{sub_{initial}} = 0.5 * \varepsilon_{sub\_max}$.

On the other hand, if there is no improvement, the BO's parameters are updated as follows:

$$NP_{cont} = NP_{cont} + 1 \text{ and } PP_{cont} = 0,$$
$$Cp = \min(0.5,\ NP_{cont} \times rcp),$$
$$\rho_{xg} = \rho_{xg\_initial}\min\left(0.5,\ \rho_{xg\_initial} + NP_{cont} \times rcp^2\right),$$
$$\text{and}$$
$$\varepsilon_{sub} = min\left(\varepsilon_{sub\_max},\ \left(\varepsilon_{sub_{initial}} - NP_{cont} \times rcp^2\right)\right).$$

The overall steps of the BO algorithm are presented in Algorithm 1.

Algorithm 1: BO

Initialize a set of random search bonobo $X_B^i = \left(X_B^1,\ X_B^2, \ldots,\ X_B^N\right)$ within the limits $X_{min}^i \leq X_B^i \leq X_{max}^i$.
Initialize the BO's parameters
Evaluate the objective function for all bonobos
Identify the alpha bonobo X_B^α
While (k < K_{max})
Determine the actual size of the temporary sub-group
Choose a bonobo using fission-fusion society strategy
Create a new bonobo X_B^{new} as follows:
if $r \leq \rho_{ph}$
Create new bonobo using promiscuous or restrictive mating strategy
else $r > \rho_{ph}$
Create new bonobo using consortship or extra-group mating strategy
end if
Calculate the objective function
Update alpha bonobo X_B^α and the BO's parameters.
$K = K + 1$
end while
Return the final best solution X_B^α

4.2. Improved Quasi-Oppositional BO (QOBO) Algorithm

As with any population-based algorithm, BO has some problems such as falling in the local optima. However, in this work, an improved BO based on three leaders' selection and quasi-opposition-based learning is developed.

4.2.1. Three Leaders

In this method, instead of using the best solution (alpha bonobo X_B^α) for updating the new bonobo X_B^{new} and ignoring the other best solutions, three leaders can be used to increase the diversity of the solutions as follows

$$X_B^\alpha = w_1 \times X_{best_1} + w_2 \times X_{best_2} + w_3 \times X_{best_3} \tag{36}$$

where

$$w_1 = \frac{r_7}{r_7 + r_8 + r_9},\ w_2 = \frac{r_8}{r_7 + r_8 + r_9},\ \text{and } w_1 = \frac{r_9}{r_7 + r_8 + r_9}$$

$r_7,\ r_8,\ and\ r_9$ are random values between [0, 1].

4.2.2. Quasi-Oppositional

Opposition-based learning (OBL) [30] has been widely used to improve many optimization techniques such as quasi-oppositional teaching-learning (QOTLBO) [31,32], Quasi-oppositional swine influenza model-based optimization with quarantine (QOSIMBO-Q) [33] and Oppositional Jaya Algorithm [34]. In the OBL, improvements can be achieved by using the candidate solution and its opposite at the same time. Hence, in this work, the opposite solution of the BO algorithm X_B^i can be expressed as presented in [35]:

$$X_B^{qnew} = C + r_{10}\left(C - \overline{X_B^{new}}\right) \tag{37}$$

where r_{10} is a random number between [0, 1], and C is a middle point between X_{min}^i and X_{max}^i which can be calculated as follows:

$$C = \frac{X_{min}^i + X_{max}^i}{2} \tag{38}$$

Additionally, $\overline{X_B^{new}}$ is the opposite solution which can be calculated as

$$\overline{X_B^{new}} = X_{min}^i + X_{max}^i - X_B^{new} \tag{39}$$

The overall steps of the improved BO based on three leaders and the quasi-oppositional method are presented in Algorithm 2.

Algorithm 2: QOBO

Initialize a set of random search bonobo $X_B^i = \left(X_B^1, X_B^2, \ldots, X_B^N\right)$ within the limits $X_{min}^i \le X_B^i \le X_{max}^i$.
Initialize the BO's parameters
Evaluate the objective function for all bonobos
Determine the alpha bonobo X_B^α using three-leader method
While (k < K_{max})
Determine the actual size of the temporary sub-group
Choose a bonobo using fission-fusion society strategy
Create a new bonobo X_B^{new} as follows:
if $r \le \rho_{ph}$
Create new bonobo using promiscuous or restrictive mating strategy
else $r > \rho_{ph}$
Create new bonobo using consortship or extra-group mating strategy
end if
Calculate the objective function for all new bonobos X_B^{new}
Find quasi-oppositional model for all new bonobos X_B^{qnew}
Calculate the objective function for all new bonobos X_B^{qnew}
if $f\left(X_B^{qnew}\right) \le f\left(X_B^{i_{new}}\right)$, $X_B^{new} = X_B^{qnew}$
Else $X_B^{new} = X_B^{new}$
end if
Update alpha bonobo X_B^α using three leader method and the BO's parameters.
$K = K + 1$
end while
Return the final best solution X_B^α

5. Case Study

To validate the robustness of the QOBO algorithm, it has been applied for addressing the studied problem of optimal configuration of the proposed multiple scenarios HRES, i.e., the PV/WT/diesel generator/BESS, PV/biomass, PV/diesel generator/BESS and WT/diesel generator/BESS. The proposed hybrid systems have been introduced in the isolated mode for satisfying the load requirements in the proposed site.

The project is applied in Aswan, Egypt as shown in Figure 4. The annual load curve over a time interval of one hour is shown in Figure 5. Figures 6–9 present solar irradiation, temperature, wind speed and atmospheric pressure in the studied region. Four standalone scenarios of the hybrid system will be evaluated for covering the load demand in that site. These configurations are: (1) PV/WT/diesel/BESS, (2) PV/biomass, (3) PV/diesel/BESS and (4) WT/diesel/BESS. The proposed QOBO is validated on optimal sizing of these four hybrid systems and the optimization results are comprehensively compared with the corresponding ones obtained from BO, HHO, AEFA and IWO algorithms.

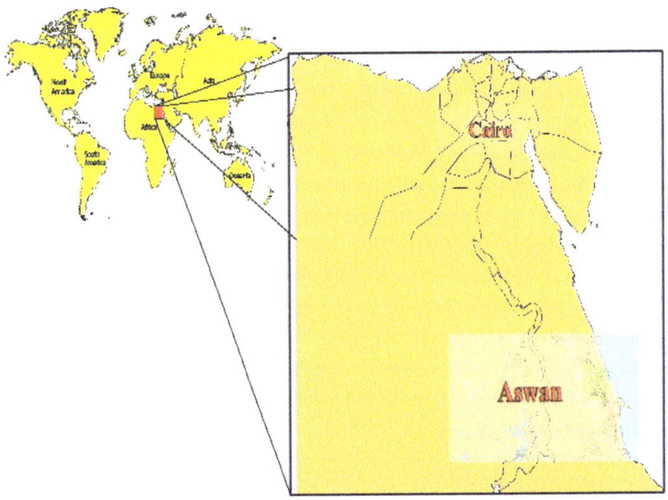

Figure 4. Location of the case study (Aswan) on the world map.

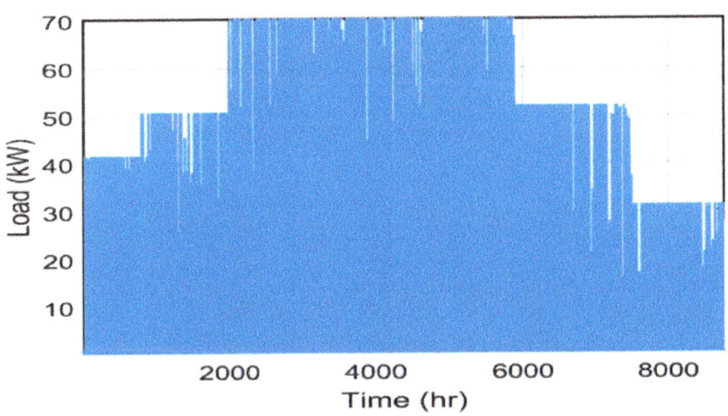

Figure 5. Annual load curve over a time interval of one hour with a peak demand of 70 kW.

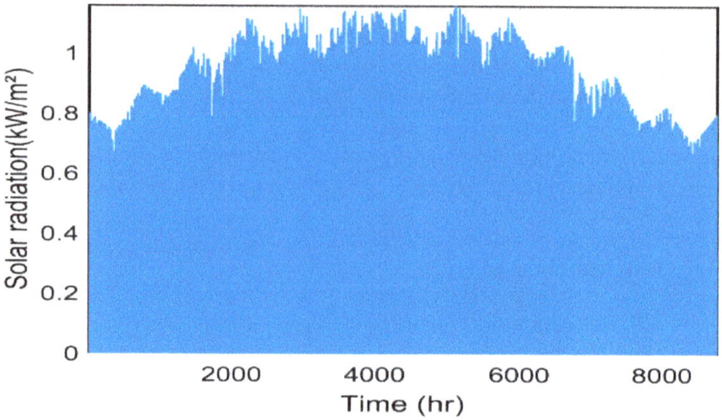

Figure 6. Solar irradiation over the studied region.

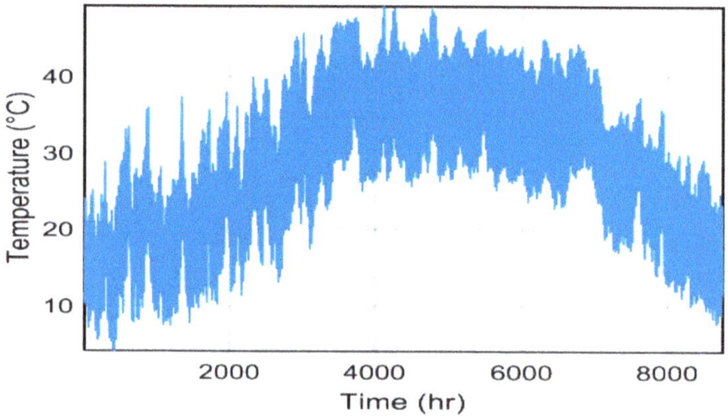

Figure 7. Temperature variation in Aswan.

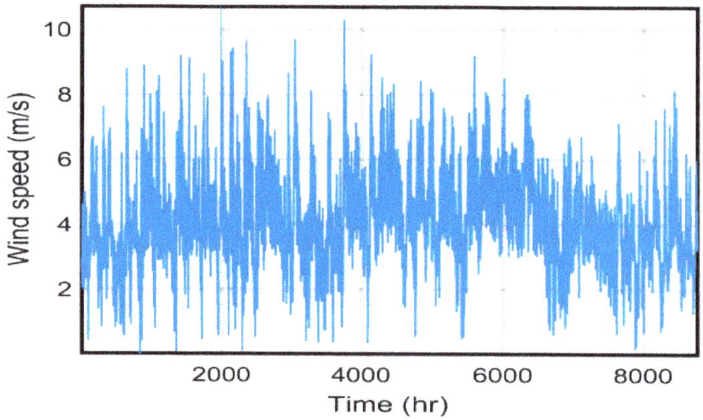

Figure 8. Annual variation of wind speed over the year in Aswan.

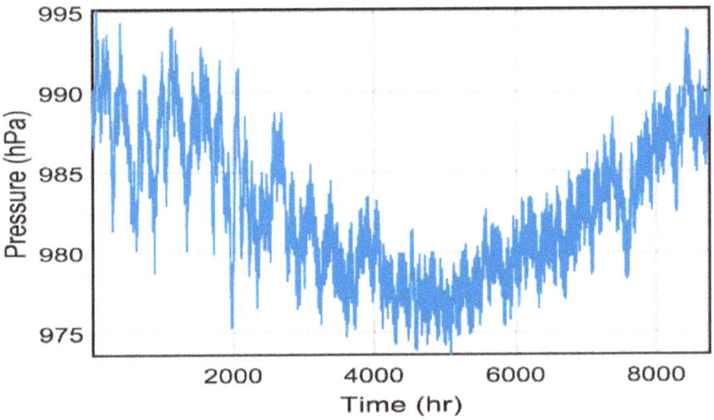

Figure 9. Atmospheric pressure variation in Aswan.

6. Results

The main object of this research paper is to find the optimal design of the proposed hybrid system and to validate the accuracy of the proposed QOBO optimization method. The optimal sizing is based on the objective functions introduced in (9) and the parameters of optimization are: (i) the area of PV system, (ii) the area swept by the WT, (iii) the rated power of diesel generator, (iv) the nominal capacity of the battery, (v) the consumption of the biomass fuel. To confirm the suitability of the QOBO in addressing such optimization problem, QOBO, BO, HHO, AEFA and IWO were launched 100 times for each configuration and statistical study was conducted based on the best minimum value of the fitness function. For a deep analysis of the obtained results and to ensure the sensitivity analysis, four indices were chosen, namely, NPC, LCOE, LPSP and the availability index. In the next subsections, the optimization results are provided for the standalone system with multiple scenarios. Modelling and simulation of the optimization problem were accomplished using MATLAB 2015a program, while the adjusting parameters for the three algorithms are the same, i.e., the number of maximum iterations is taken as 100 iterations and the search agents' number is 30 agents. The input technical and economic data for the system components are presented in Table 1. The results of the statistical measurements for the modified QOBO and the conventional BO with HHO, AEFA and IWO algorithms are listed in Tables 2 and 3. From the previously mentioned tables, the reader can conclude that the QOBO technique generates the best minimum value of the fitness function in all cases. The convergence curves of the 100 iterations implemented for all the studied configurations using QOBO, BO, HHO, AEFA and IWO are presented in Figure 10a–d.

Table 1. Units for magnetic properties.

Symbol	Quantity	Conversion
N	Project lifetime	20 years
i_r	Interest rate	13.25%
μ	Escalation rate	2%
δ	Inflation rate	12.27%
	PV system	
λ_{pv}	PV initial cost	300 $/m^2
θ_{pv}	Annual O&M cost of PV	$0.01 * \lambda_{pv}$ $/m^2/year
η_r	Reference efficiency of the PV	25%
η_t	Efficiency of MPPT	100%
T_r	PV cell reference temperature	25 °C
β	Temperature coefficient	0.005 °C
$NOCT$	Nominal operating cell temperature	47 °C
N_{pv}	PV system lifetime	20 years
	WT system	
λ_{wind}	Wind initial cost	125 $/m^2
θ_{wind}	Annual O&M cost of wind	$0.01 * \lambda_{wind}$ $/m^2/year
C_{p_wind}	Maximum power coefficient	48%
V_{ci}	Cut-in wind speed	2.6 m/s
V_{co}	Cut-out wind speed	25 m/s
V_r	Rated wind speed	9.5 m/s
N_{wind}	Wind system lifetime	20 years
	Diesel generator	
λ_{dg}	Diesel initial cost	250 $/kW
θ_{dg}	Annual O&M cost of diesel	0.05 $/h
R_{dg}	Replacement cost	210 $/kW
p_f	Fuel price in Egypt	0.43 $/L
N_{diesel}	Diesel system lifetime	7 years
	BESS	
λ_{bat}	Battery initial cost	100 $/kWh
θ_{bat}	Annual operation and maintenance cost of battery	$0.03 * \lambda_{bat}$ $/m^2/year
DOD	Depth of discharge	80%
η_b	Battery efficiency	97%
SOC_{min}	Minimum state of charge	20%
SOC_{max}	Maximum state of charge	80%
N_{bat}	Battery system lifetime	5 years
	Inverter	
λ_{inv}	Inverter initial cost	400 $/m^2
θ_{inv}	Annual O&M cost of inverter	20 $/year
η_{inv}	Inverter efficiency	97%

6.1. Validation of QOBO Algorithm

The results of the statistical measurements for the modified QOBO, the conventional BO, HHO, AEFA and IWO algorithms are listed in Tables 2 and 3. From Table 2, the reader can find the results of the optimal sizing for the multiple scenarios studied, as well as the convergence time of each simulation, and conclude that the QOBO algorithm finds the best results with a short time compared with the other algorithms. From Table 3, the reader can compare between algorithms and the different scenarios of the proposed hybrid system using multiple factors. Briefly, it is noticed that the hybrid PV/biomass system is highly competitive, mainly using the developed QOBO algorithm, the optimized system is calculated with $110,807, which means an LCOE of 0.1053 $/kWh, the constraints are satisfied and the project is 100% supplied by renewable energy sources. In this scenario, the performances of the QOBO and the BO are almost equal, while in other scenarios, the difference is clearly noticed.

Table 2. Sizing results of different scenarios obtained from different optimization methods.

Hybrid Power System	Algorithm	PV (m^2)	Wind (m^2)	Diesel (kW)	Battery (kWh)	Biomass (t/year)	Time(s)
PV/WT/Diesel/BESS	QOBO	484.765	0	1.2142	13.4390	//	51,507
	BO	248.002	998.505	0.6480	14.8052	//	164,242
	HHO	513.105	305.293	0.5204	14.6552	//	30,655
	AEFA	329.159	176.277	5.4696	18.6552	//	10,531
	IWO	830.791	136.557	10.296	5.8224	//	57,938
PV/Biomass	QOBO	293.971	//	//	//	1020.18	32,104
	BO	293.972	//	//	//	1020.31	122,417
	HHO	298.860	//	//	//	2040.47	10,453
	AEFA	302.980	//	//	//	1185.76	3855
	IWO	365.515	//	//	//	2739.00	36,098
PV/Diesel/BESS	QOBO	376.011	//	1.3402	58.9083	//	16,799
	BO	336.253	//	2.9170	52.1928	//	33,009
	HHO	482.756	//	1.7843	13.7590	//	13,983
	AEFA	386.692	//	1.6713	55.7583	//	6237
	IWO	748.387	//	4.0111	51.4565	//	24,630
WT/Diesel/BESS	QOBO	//	2726.29	91.141	72.375	//	26,510
	BO	//	2823.34	42.637	72.371	//	66,514
	HHO	//	2808.76	74.565	73.230	//	135,097
	AEFA	//	3015.08	72.963	72.653	//	78,697
	IWO	//	4318.76	78.218	82.7987	//	26,960

Table 3. Factor results for all scenarios.

Hybrid Power System	Algorithm	NPC ($)	LCOE ($/kWh)	LPSP (%)	Availability (%)	Renewable Energy (%)	Battery Daily Autonomy (day)
PV/WT/Diesel/BESS	QOBO	175,651	0.1669	0.019	98.87	98.15	0.5826
	BO	209,096	0.1986	0.050	96.99	99.75	0.6418
	HHO	201,109	0.1910	0.025	99.23	99.88	0.6353
	AEFA	183,284	0.1741	0.026	99.33	96.88	0.8087
	IWO	347,523	0.3301	0.014	99.68	97.72	0.2524
PV/Biomass	QOBO	110,807	0.1053	0.050	96.03	100	//
	BO	110,808	0.1053	0.050	96.03	100	//
	HHO	114,098	0.1084	0.046	96.94	100	//
	AEFA	113,410	0.1077	0.040	96.93	100	//
	IWO	130,491	0.1240	0.018	98.70	100	//
PV/Diesel/BESS	QOBO	153,401	0.1457	0.049	98.63	97.25	2.5536
	BO	167,981	0.1596	0.050	98.72	92.88	2.2625
	HHO	183,501	0.1743	0.017	98.94	97.27	0.5964
	AEFA	160,774	0.1527	0.042	98.74	96.70	2.4171
	IWO	287,730	0.2733	0.026	99.16	96.12	2.2306
WT/Diesel/BESS	QOBO	1,095,270	1.0405	0.014	99.85	70.03	3.9509
	BO	1,098,685	1.0437	0.003	99.97	71.3527	1.8483
	HHO	1,123,579	1.0673	0.008	99.92	70.2407	3.1745
	AEFA	1,119,533	1.0635	0.008	99.92	73.6967	3.1494
	IWO	1,319,108	1.2531	0.008	99.92	81.8292	3.3907

6.2. Combinations of the Studied System Components

In this section, the results obtained in the convergence simulation of the NPC as a fitness function using the QOBO are presented. The optimized parameter results (i.e., A_{pv}, A_{wind}, P_{dgn}, P_{Cap_bat}, P_{BM}) for all suggested combinations are listed in Table 3 with the rating of the inverter that takes the value of the peak load demand. From Figure 10, the reader can notice that using QOBO, BO, HHO, AEFA and IWO algorithms, the best minimum values of fitness function (NPC) is obtained for the second configuration, i.e., hybrid PV/biomass energy system. From the table, it is obvious that QOBO generates the minimum value of LCOE in all cases.

The reliability of the proposed scenarios of the proposed HRES are respected and the availability of power is highly assured, the penetration RES is considered in this paper, while different results are obtained. The minimum penetration of 70% is obtained for the WT/Diesel/Battery scenario while the maximum penetration of 99.75% is obtained for the PV/WT/Diesel/BESS scenario. The daily battery autonomy is also influenced by the configuration of the HRES, the best autonomy is achieved for the WT/Diesel/BESS scenario taking nearly 4 days, while the minimum autonomy is obtained in PV/WT/Diesel/BESS case with only 6 h. The last system is composed of the different energy resource which explains the independence for a specific resource. Table 4 presents a detailed overview of all costs needed, for all scenarios presented and for all proposed algorithms.

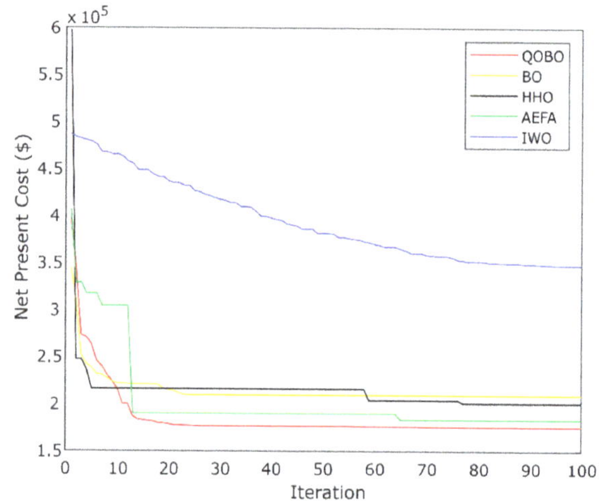

(a)

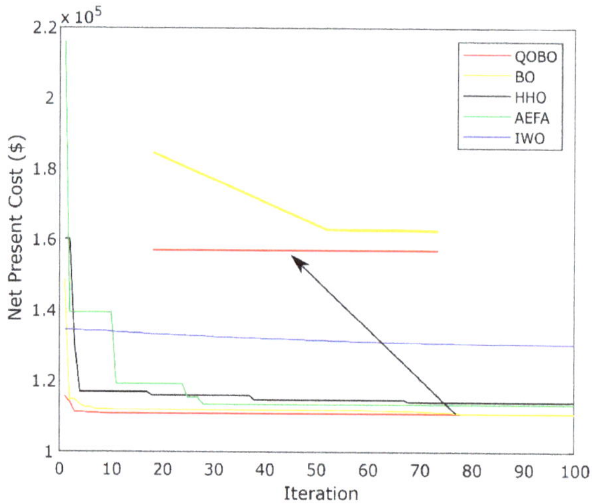

(b)

Figure 10. *Cont.*

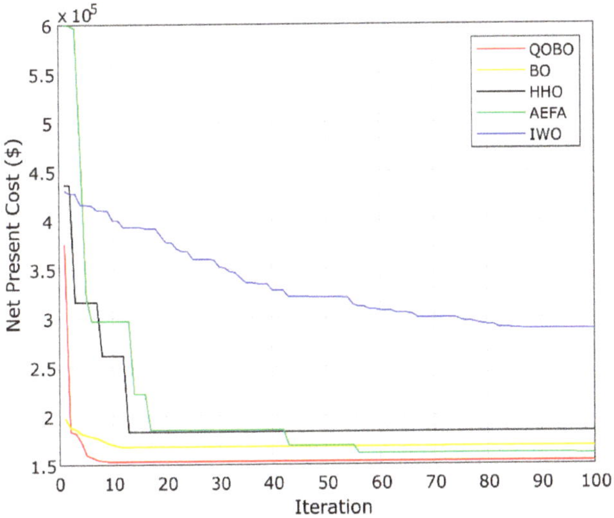

(c)

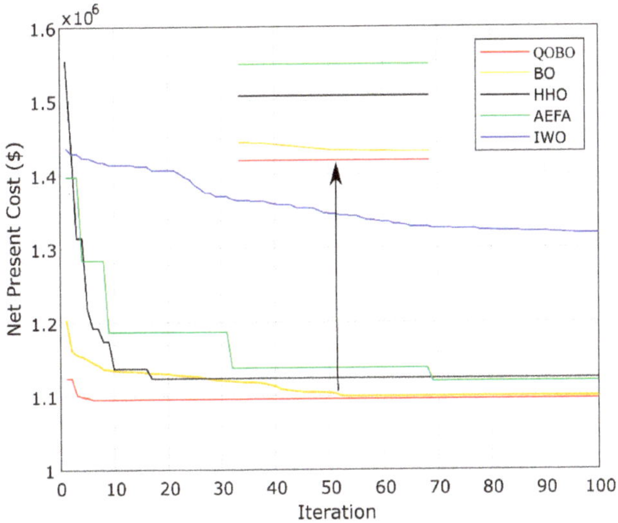

(d)

Figure 10. Convergence of the objective function of all algorithms for different scenarios; (**a**) PV/WT/Diesel/BESS, (**b**) PV/Biomass, (**c**) PV/Diesel/BESS, (**d**) WT/Diesel/BESS.

Table 4. Convergence of objective function of different scenarios.

Hybrid Power System	Algorithm	PV			Wind			Diesel				Battery		Inverter		Biomass		
Costs		Inv	O&M	Rep	Inv	O&M	Rep	Inv	O&M	Rep	Fuel	Inv	Rep	Inv	Rep	Inv	O&M	Fuel
Scenario I	QOBO	145,429	11,558	0	0	0	0	303	1792	465	17,080	1343	90	28,400	158		//	
	BO	74,400	5913	0	124,813	9920	0	162	526	248	5244	1480	99	28,400	158		//	
	HHO	153,931	12,234	0	38,161	3033	0	130	330	199	3729	1465	98	28,400	158		//	
	AEFA	98,747	7848	0	22,034	1751	0	1367	537	2097	44,542	1865	125	28,400	158.		//	
	IWO	249,237	19,809	0	17,069	1356	0	2574	470	3949	80,625	582	39	28,400	158		//	
Scenario II	QOBO	88,191	7009	0	//	//	//	//	//	//	//	//	//	28,400	158	1040	103	4696
	BO	88,191	7009	0	//	//	//	//	//	//	//	//	//	28,400	158	1040	103	4696
	HHO	89,658	7126	0	//	//	//	//	//	//	//	//	//	28,400	158	2080	129	5907
	AEFA	90,894	7224	0	//	//	//	//	//	//	//	//	//	28,400	158	1208	105	4820
	IWO	109,654	8715	0	//	//	//	//	//	//	//	//	//	28,400	158	2792	97	4416
Scenario III	QOBO	112,803	8965	0	//	//	//	335	1869	514	19,339	5890	395	28,400	158		//	
	BO	100,875	8017	0	//	//	//	729	1994	1118	43,812	5219	350	28,400	158		//	
	HHO	144,826	11,510	0	//	//	//	446	1792	684	25,102	1375	92	28,400	158		//	
	AEFA	116,007	9220	0	//	//	//	417	1855	641	24,008	5575	374	28,400	158		//	
	IWO	224,516	17,844	0	//	//	//	1002	1756	1538	55,745	5145	345	28,400	158		//	
Scenario IV	QOBO	//	//	//	340,787	27,085	0	18,093	921	27,759	720,800	9114	612	28,400	158		//	
	BO	//	//	//	352,917	28,050	0	18,092	912	27,757	717,633	4263	286	28,400	158		//	
	HHO	//	//	//	351,094	27,905	0	18,641	915	28,599	740,365	7323	491	28,400	158		//	
	AEFA	//	//	//	376,885	29,955	0	18,240	887	27,985	714,885	7265	488	28,400	158		//	
	IWO	//	//	//	539,845	42,907	0	20,699	774	31,757	766,838	7821	525	28,400	158		//	

6.3. Sensitivity Analysis

RES is intermittent which can be affected by any variation of sizing, meteorological or economic data. The sensitivity analysis is a method that helps to select and/or to expect the optimal configuration of the hybrid system. The sensitivity analysis in this paper is implemented on the best scenario of the proposed, i.e., the PV/Biomass in the Aswan region. The selection of the sensitivity variables is based on the sizing of components in order to analyze the effect of sizing variation on four factors which are NPC, LCOE, LPSP and the Availability index.

Figure 11 shows the effect of variation in the sizing of PV and biomass units on the NPC. The PV sizing is highly impacted by the total cost of the hybrid PV/Biomass system, which means that in the case of reducing the area of PV units the NPC is reduced too. On the other hand, if the area covered by PV modules is increased, the NPC increases too. The variation in the sizing of biomass unit is increased throughout the interval −20 to 20 slowly and it has no noticeable impact on the NPC anyway. Figure 12 shows the effect of variation of PV and biomass sizing on the LCOE. The NPC and the LCOE are linked with a linear equation which means that they have the same shape. The LCOE reached 0.08 $/kWh when the area of the PV system is reduced by 20%. Figure 13 shows the impact of variation in the sizing of PV and biomass systems on the LPSP. The impact of PV size is very important for the LPSP, because when the size of the PV system is increased the LPSP is enhanced, mainly in the −20% to 0% interval. When the PV size is changed in the interval of 0% to +20%, the LPSP is increased to 2% while when the PV size is changed to −20%, the change in LPSP equals 16.4% which is a very bad sign for system building. The Biomass system does not affect the value of the LPSP and the transition between −4% to 0% is explained as the obtained sizing of the system is optimum. Figure 14 shows the impact of the variation of PV and Biomass sizing on the availability index. The availability index enhanced exponentially with the increase in the PV sizing. In the interval between −20% and 0, availability progresses quickly, while after zero, the availability begins to be stabilized and it is clearly shown in the interval between +12% and +20%.

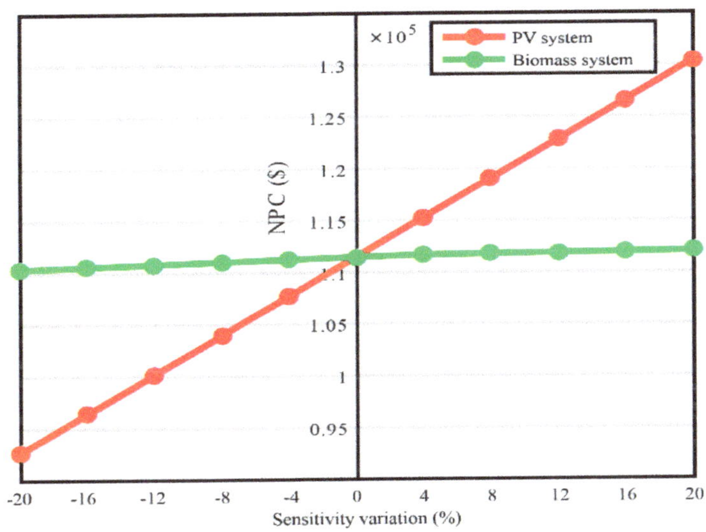

Figure 11. Sensitivity analysis application for net present cost (NPC).

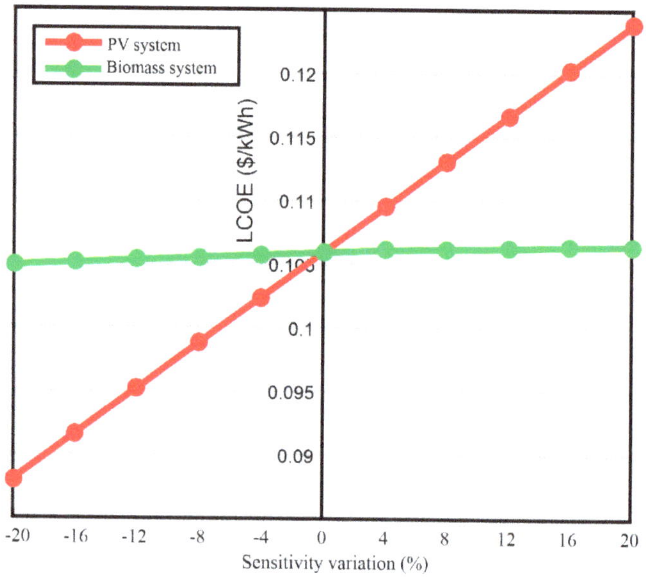

Figure 12. Sensitivity analysis application for Levelized Cost of Energy (LCOE).

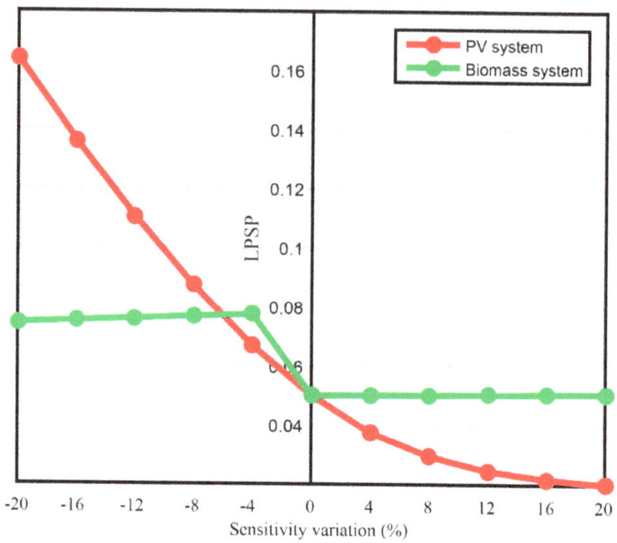

Figure 13. Sensitivity analysis application for Loss of Power Supply Probability (LPSP).

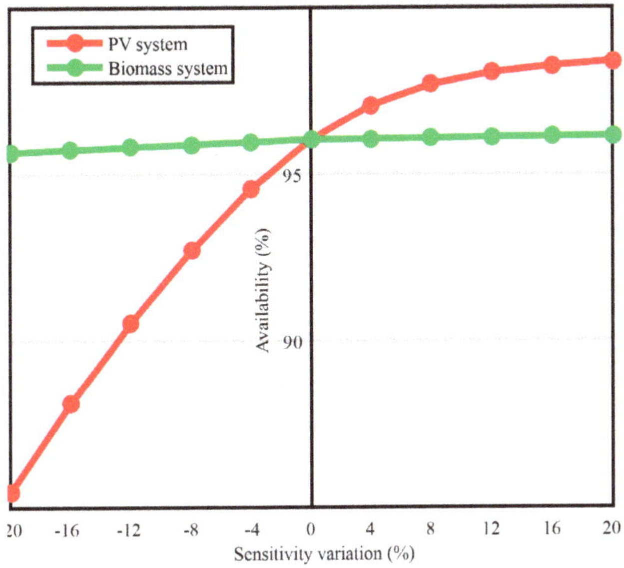

Figure 14. Sensitivity analysis application for the Availability index.

The PV system through this analysis is demonstrated as a very important element for having a high hybrid systems criterion. However, the Biomass system helps the PV to satisfy the constraints and its variation does not have a serious impact on the performance of the hybrid system.

7. Conclusions

With the increased penetration level of RES into electrical energy production in the microgrid systems, new challenges have emerged on the international scene. These challenges are represented in finding ways to optimize the design of the hybrid system by using smart algorithms and software. Among these dilemmas, the economic cost and feasibility of installing systems in different locations in the world is considered the most important challenge. Therefore, this research proposes a developed algorithm called Quasi-Oppositional Bonobo Optimizer (QOBO) for the optimal economic design of a stand-alone hybrid microgrid system in Aswan, Egypt. Four configurations of the hybrid system have been implemented, which consist of RES (PV panels, WT and biomass) with diesel generators and battery storage systems. The obtained results showed that the PV/Biomass scenario is the most cost-effective system with an NPC of $110,807 and LCOE of 0.1053 $/kWh; otherwise, the best configuration of the microgrid system contained 293.971 m^2 of PV and 1020.18 ton/year consumed by the biomass system; the PV/Diesel/BESS scenario is also cost-effective with NPC of $153,401 and LCOE of 0.1457 $/kWh. On the other side, the LPSP and availability index are satisfied and without the need for traditional resources. Additionally, the results showed the ability of the QOBO algorithm to reach the optimal solution in a shorter time and with better efficiency compared to the traditional BO, HHO, AEFA and IWO algorithms in all cases studies. Furthermore, a sensitivity analysis of the proposed systems scenarios was performed to obtain the impact of unit size on the performance of the hybrid system, where it has been emphasized that PV system sizing is very important and has a great impact on the overall performance of the system. The obtained results from this study would be useful material for decision makers working on the development of the renewable energy sector in Egypt. In future studies, it is suggested to apply the proposed QOBO in other engineering problems.

Author Contributions: Conceptualization, S.K. and M.K.; Data curation, O.H.M., H.M.S. and A.S.; Formal analysis, F.J., M.A. and H.M.S.; Methodology, M.K., A.S. and S.K.; Resources, O.H.M., S.K. and H.M.S.; Software,

A.S., M.K.; Supervision, F.J.; Validation, H.M.S. and O.H.M.; Visualization, S.K., A.S. and M.A.; Writing—original draft, M.K., O.H.M., A.S. and H.M.S.; Writing—review & editing, S.K., M.A. and F.J. All authors together organized and refined the manuscript in the present form. All authors have approved the final version of the submitted paper. All authors have read and agreed to the published version of the manuscript.

Funding: This work was supported in part by the NSFC, China-ASRT, Egypt, Joint Research Fund, under Grant 51861145406.

Acknowledgments: The authors gratefully acknowledge the contribution of the NSFC (China)-ASRT (Egypt) Joint Research Fund, Project No. 51861145406 for providing partial research funding to the work reported in this research.

Conflicts of Interest: The authors declare no conflict of interest.

Nomenclature

Symbols			
A	Availability index	η_b	Efficiency of the battery (%)
A_g	Coefficient of consumption curve (a = 0.246 L/kW)	η_{bio}	Efficiency of the biomass system (%)
AD	Daily autonomy of the battery (day)	η_{inv}	Efficiency of the inverter (%)
A_{pv}	Area covered by PV panels (m^2)	η_{pv}	Efficiency of the PV system (%)
A_{tt}	Cross-sectional area of the tidal (m^2)	η_r	Reference efficiency of PV panels (%)
A_{wind}	Swept area by the wind turbine (m^2)		
C	Capital Cost ($)	P_{wind}	Output power of the wind turbine (kW)
$C_{Battery}$	Capacity of the Battery (kWh)	R	Replacement Cost ($)
C_p	Maximum power coefficient (%)	T	Temperature (°C)
CV_{bio}	Calorific value of the organic material (MJ/kg)	T_a	Ambient temperature (°C)
DOD	Depth of Discharge (%)	$Total_{av}$	Total biomass available (t/yr)
E_l	Load demand (kWh)	T_r	Reference temperature of solar cell (°C)
F_{dg}	Fuel consumption of the diesel generator (L/h)	V	Wind speed (m/s)
FC_{dg}	Fuel Cost for one year ($/Year)	V_{ci}	Cut-in wind speed (m/s)
I	Solar irradiation (kW/m^2)	V_{co}	Cut-out wind speed (m/s)
i_r	Interest rate (%)	V_r	Rated wind speed (m/s)
N	project lifetime (year)	B_g	Coefficient of consumption curve (b = 0.08415 L/kW)
$NOCT$	Nominal operating cell temperature (°C)	η_t	Efficiency MPPT system (%)
NPC	Net Present Cost ($)	β	Temperature coefficient (0.004 to 0.006 °C)
OM	Maintenance and Operation ($)	ρ	Air density (Kg/m^3)
P_{dg}	Rated power of the diesel generator (kW)	λ_{bat}	Initial cost of the battery system ($/kWh)
P_f	Fuel price ($/L)	λ_{bg}	Initial cost of biomass system ($/kW)
P_{bg}	Generated power of the biogas plant (kW)	λ_{dg}	Initial cost of diesel generator ($/kW)
P_{BM}	Biomass power (kW)	$\lambda_{PV,WT}$	Initial cost of PV and WT ($/m^2)
P_{pv}	Output power of the PV (kW)	δ	Inflation rate (%)
P_r	Rated power (kW)	μ	Escalation rate (%)
P_{re}	Power from renewable energy systems	θ_1	Biomass annual fixed O&M cost ($/kW/year)
P_w	Annual working of biomass (kWh/Year)	θ_2	Biomass variable O&M cost ($/kW h)
Acronyms			
AEFA	Artificial Electric Field Algorithm	HSA	Harmony Search Algorithm
ACS	Annualized cost of the system	IWO	Invasive Weed optimization Algorithm
BESS	Battery Energy Storage System	LCOE	Levelized Cost of Energy
BO	Bonobo Optimizer Algorithm	LPSP	Loss of Power Supply Probability
BOQO	Improved Quasi Oppositional BO Algorithm	MOPSO	Multiple Objective Particle Swarm Optimization
COE	Cost of Energy	NPC	Net present cost
CRF	Capital Recovery Factor	PSO	Particle Swarm Optimization
HOMER	Hybrid Optimization of Multiple Energy Resources	PV	Photovoltaic
HRES	Hybrid Renewable Energy Systems	RF	Renewable Fraction
HHO	Harris Hawks Optimization	WT	Wind Turbine

Appendix A. Algorithms

Appendix A.1. Harris Hawks Optimization Algorithm

Heidari and et al. [36] proposed a new nature-inspired optimization algorithm called Harris Hawks Optimizer. They were inspired by the cooperative behavior and chasing style of Harris hawks. The modeling of this technique is based firstly on an exploration phase; afterwards, the transition from exploration to exploitation, then the exploitation phase and, finally, the soft besiege. The modeling is taken on for all strategies for exploring a prey, surprise pounce and different attacking methods of Harris hawks. The pseudo-code of the HHO algorithm is proposed below.

Algorithm A1: Pseudo code of HHO

Initialize the population size and max iteration (K_{max})
Initialize a set random rabbit location, within the limits $X^{i}_{min} \leq X^{i}_{rabbit} \leq X^{i}_{max}$.
Evaluate the objective function for all rabbits
While (k < K_{max})
Calculate the fitness of hawks
Set x_{rabbit} in the best location
for each hawk do
Update the initial energy E_0, energy E and jump strength J;
$E_0 = 2\text{rand}() - 1$, $E = 2E_0\left(1 - \frac{t}{T}\right)$, J = 2(1 − rand ())
if (|E| ≥ 1) then
Exploration phase
if (|E| < 1) then
Exploitation phase
if (r ≥ 0.5 and |E| ≥ 0.5) then
Soft besiege
else if (r ≥ 0.5 and |E| < 0.5) then
Hard besiege
else if (r < 0.5 and |E| ≥ 0.5) then
Soft besiege with progressive rapid dives
else if (r < 0.5 and |E| < 0.5) then
Hard besiege with progressive rapid dives
Return x_{rabbit}

Appendix A.2. Artificial Electric Field Algorithm

Anita and Yadav [37] were inspired by Coulomb's law of electrostatic force to create a novel artificial electric field algorithm. The concepts of electric field and charged particles provide us a strong theory for the working force of attraction or repulsion between two charged particles. The pseudo code of the AEFA algorithm is proposed in Algorithm A2.

Algorithm A2: Pseudo code of AEFA

Initialize a set of random population $X^{i}_{B} = \left(X^{1}_{B}, X^{2}_{B}, \ldots, X^{N}_{B}\right)$ of N size, within the limits
$X^{i}_{min} \leq X^{i}_{B} \leq X^{i}_{max}$.
Initialize the velocity to a random value
Evaluate the fitness of whole population
Set the iteration to zero
Reproduction and Updating
While criteria not satisfied do
Calculate K (t), best (t) and worst (t)
for i = 1: N do
Evaluate the fitness values
Calculate the total force in each direction
Calculate the acceleration
$V_i(t+1) = \text{rand}() \times V_i(t) + a_i(t)$
$X_i(t+1) = X_i(t) + V_i(t+1)$
end for
end while

Appendix A.2.1. Invasive Weed Optimization Algorithm

Invasive weed optimization is a numerical stochastic optimization algorithm inspired by colonizing weeds, which was introduced in 2006 by Mehrabian and Lucas [38]. In IWO, a certain number of weeds make up the whole population, and each weed comprises a set of decision variables. Weeds are a serious threat to desirable plants because they are plants that are invasive and hardy.

Weeds are plants which are vigorous and invasive; they pose a serious threat to desirable, cultivated plants in agriculture. Weeds have shown to be very robust and adaptive to change in the environment. The IWO optimization algorithm has been modeled based on four steps: initialization, reproduction, spatial dispersal and competitive exclusion.

- *Initialization and Production*

Firstly, the population is spread over the research space randomly; afterwards, each plant is allowed to produce seeds depending on its own fitness; the production of seeds is not only allowed for the better plants' fitness as in the other evolutionary algorithms, but the reproduction step of IWO is also proposed to give a chance to infeasible individuals to survive and reproduce similar to the mechanism which occurs in nature. The weeds producing seeds can be formulated as follows:

$$Weed_n = \frac{f - f_{min}}{f_{max} - f_{min}}(s_{max} - s_{min}) + s_{min} \quad \text{(A1)}$$

where in each iteration, f is the current weed's fitness. f_{max} and f_{min} represent the max and min fitness values, respectively. s_{max} and s_{min} represent the max and min values of the weeds, respectively.

- *Spatial Dispersal*

The generated seeds are being randomly distributed over the search space such that they abode near the parent plant. However, the standard deviation (σ) of the random function will be reduced in every iteration, the nonlinear alteration equation of the standard deviation at each iteration is presented as follows:

$$\sigma_{inter} = \frac{(iter_{max} - iter)^n}{(iter_{max})^n}\left(\sigma_{initial} - \sigma_{final}\right) + \sigma_{final} \quad \text{(A2)}$$

where $iter_{max}$ is the maximum iteration, n is the nonlinear modulation index, $\sigma_{initial}$ and σ_{final} are the initial and final values of standard deviation, respectively.

- *Competitive Exclusion*

In a colony, the maximum number allowed of plants is limited; for that, competitive exclusion is applied. The plant that leave no offspring would go extinct; otherwise, they can survive. After some iterations, the number of plants in a colony will reach its maximum through the reproduction step, the seeds and their parents are ranked together, and all plants in the research space are considered as weeds; afterwards, weeds with lower fitness are eliminated.

The overall steps of the IWO algorithm are presented in Algorithm A3.

Algorithm A3: Pseudo code of IWO

Initialize a set of random weeds, $weed^i_B = \left(weed^1_B, weed^2_B, \ldots, weed^N_B\right)$ within the limits $weed^i_{min} \le weed^i_B \le weed^i_{max}$.
Set the IWO's parameters
Evaluate the objective function for all weeds
While $(iter < iter_{max})$
Calculate the best and worst fitness in the colony
Calculate the σ
for each weed in the colony
Calculate the number of seeds following the fitness of each weed
Add the seeds to their parents in the colony
if $Size_{max} \le Nb_{population}$
Sort the new population according to their fitness
Eliminate the worst fitness in order to achieve the $Size_{max}$ allowed
end if
end for
Update iteration $iter = iter + 1$
end while
Return the final best solution

References

1. Kharrich, M.; Mohammed, O.H.; Mohammed, Y.S.; Akherraz, M. A Review on Recent Sizing Methodologies for Hybrid Microgrid Systems. *Int. J. Energy Convers.* **2019**, *7*, 230–240. [CrossRef]
2. Babatunde, O.M.; Munda, J.L.; Hamam, Y. A Comprehensive State-of-the-Art Survey on Hybrid Renewable Energy System Operations and Planning. *IEEE Access* **2020**, *8*, 75313–75346. [CrossRef]
3. Khatib, H. IEA world energy outlook 2011—A comment. *Energy Policy* **2012**, *48*, 737–743. [CrossRef]
4. Zhang, W.; Maleki, A.; Rosen, M.A.; Liu, J. Optimization with a simulated annealing algorithm of a hybrid system for renewable energy including battery and hydrogen storage. *Energy* **2018**, *163*, 191–207. [CrossRef]
5. Diab, A.A.Z.; Sultan, H.M.; Mohamed, I.S.; Kuznetsov, O.N.; Do, T.D. Application of different optimization algorithms for optimal sizing of PV/wind/diesel/battery storage stand-alone hybrid microgrid. *IEEE Access* **2019**, *7*, 119223–119245. [CrossRef]
6. Diab, A.A.Z.; Sultan, H.M.; Kuznetsov, O.N. Optimal sizing of hybrid solar/wind/hydroelectric pumped storage energy system in Egypt based on different meta-heuristic techniques. *Environ. Sci. Pollut. Res.* **2019**, *27*, 1–23. [CrossRef]
7. Mohammed, O.; Amirat, Y.; Benbouzid, M.; Feld, G. Optimal Design and Energy Management of a Hybrid Power Generation System Based on Wind/Tidal/Pv Sources: Case Study for the Ouessant French Island. In *Smart Energy Grid Design for Island Countries*; Springer: Berlin/Heidelberg, Germany, 2017; pp. 381–413.
8. Lian, J.; Zhang, Y.; Ma, C.; Yang, Y.; Chaima, E. A review on recent sizing methodologies of hybrid renewable energy systems. *Energy Convers. Manag.* **2019**, *199*, 112027. [CrossRef]
9. Yadav, D.K.; Girimaji, S.P.; Bhatti, T.S. Optimal Hybrid Power System Design Using HOMER. In Proceedings of the 2012 IEEE 5th India International Conference on Power Electronics (IICPE), Delhi, India, 6–8 December 2012; pp. 1–6.
10. Mohammed, O.H. Optimal Hybridisation of a Renewable System to Fulfill Residential Electrical Load: In Mosul, Iraq. In Proceedings of the 2018 Third Scientific Conference of Electrical Engineering (SCEE), Baghdad, Iraq, 9–20 December 2018; IEEE: Piscataway, NJ, USA, 2018; pp. 209–213.
11. Papaefthymiou, S.V.; Papathanassiou, S.A. Optimum sizing of wind-pumped-storage hybrid power stations in island systems. *Renew. Energy* **2014**, *64*, 187–196. [CrossRef]
12. Hazem Mohammed, O.; Amirat, Y.; Benbouzid, M. Economical evaluation and optimal energy management of a stand-alone hybrid energy system handling in genetic algorithm strategies. *Electronics* **2018**, *7*, 233. [CrossRef]
13. Rezk, H.; Al-Dhaifallah, M.; Hassan, Y.B.; Ziedan, H.A. Optimization and Energy Management of Hybrid Photovoltaic-Diesel-Battery System to Pump and Desalinate Water at Isolated Regions. *IEEE Access* **2020**, *8*, 102512–102529. [CrossRef]
14. Wang, R.; Xiong, J.; He, M.-F.; Gao, L.; Wang, L. Multi-objective optimal design of hybrid renewable energy system under multiple scenarios. *Renew. Energy* **2020**, *151*, 226–237. [CrossRef]
15. Nandi, S.K.; Ghosh, H.R. Prospect of wind–PV-battery hybrid power system as an alternative to grid extension in Bangladesh. *Energy* **2010**, *35*, 3040–3047. [CrossRef]
16. Khan, M.; Iqbal, M. Pre-feasibility study of stand-alone hybrid energy systems for applications in Newfoundland. *Renew. Energy* **2005**, *30*, 835–854. [CrossRef]
17. Sen, R.; Bhattacharyya, S.C. Off-grid electricity generation with renewable energy technologies in India: An application of HOMER. *Renew. Energy* **2014**, *62*, 388–398. [CrossRef]
18. Hemeida, A.; El-Ahmar, M.; El-Sayed, A.; Hasanien, H.M.; Alkhalaf, S.; Esmail, M.; Senjyu, T. Optimum design of hybrid wind/PV energy system for remote area. *Ain Shams Eng. J.* **2020**, *11*, 11–23. [CrossRef]
19. Xu, X.; Hu, W.; Cao, D.; Huang, Q.; Chen, C.; Chen, Z. Optimized sizing of a standalone PV-wind-hydropower station with pumped-storage installation hybrid energy system. *Renew. Energy* **2020**, *147*, 1418–1431. [CrossRef]
20. Barbaro, M.; Castro, R. Design optimisation for a hybrid renewable microgrid: Application to the case of Faial island, Azores archipelago. *Renew. Energy* **2020**, *151*, 434–445. [CrossRef]
21. Alshammari, N.; Asumadu, J. Optimum Unit Sizing of Hybrid Renewable Energy System Utilizing Harmony Search, Jaya and Particle Swarm Optimization Algorithms. *Sustain. Cities Soc.* **2020**, *60*, 102255. [CrossRef]

22. Das, A.K.; Pratihar, D.K. A New Bonobo Optimizer (BO) for Real-Parameter Optimization. In Proceedings of the 2019 IEEE Region 10 Symposium (TENSYMP), Kolkata, India, 7–9 June 2019; IEEE: Piscataway, NJ, USA, 2019; pp. 108–113.
23. Kharrich, M.; Mohammed, O.H.; Akherraz, M. Assessment of Renewable Energy Sources in Morocco using Economical Feasibility Technique. *Int. J. Renew. Energy Res.* **2019**, *9*, 1856–1864.
24. Kharrich, M.; Mohammed, O.H.; Akherraz, M. Design of Hybrid Microgrid PV/Wind/Diesel/Battery System: Case Study for Rabat and Baghdad. *EAI Endorsed Trans. Energy Web* **2020**, *7*. [CrossRef]
25. Ramli, M.A.M.; Bouchekara, H.R.E.H.; Alghamdi, A.S. Optimal sizing of PV/wind/diesel hybrid microgrid system using multi-objective self-adaptive differential evolution algorithm. *Renew. Energy* **2018**, *121*, 400–411. [CrossRef]
26. Ghiasi, M. Detailed study, multi-objective optimization, and design of an AC-DC smart microgrid with hybrid renewable energy resources. *Energy* **2019**, *169*, 496–507. [CrossRef]
27. Movahediyan, Z.; Askarzadeh, A. Multi-objective optimization framework of a photovoltaic-diesel generator hybrid energy system considering operating reserve. *Sustain. Cities Soc.* **2018**, *41*, 1–12. [CrossRef]
28. Heydari, A.; Askarzadeh, A. Optimization of a biomass-based photovoltaic power plant for an off-grid application subject to loss of power supply probability concept. *Appl. Energy* **2016**, *165*, 601–611. [CrossRef]
29. Sultan, H.M.; Menesy, A.S.; Kamel, S.; Tostado-Véliz, M.; Jurado, F. Parameter Identification of Proton Exchange Membrane Fuel Cell Stacks Using Bonobo Optimizer. In Proceedings of the 2020 IEEE International Conference on Environment and Electrical Engineering and 2020 IEEE Industrial and Commercial Power Systems Europe (EEEIC/I&CPS Europe), Madrid, Spain, 9–12 June 2020; IEEE: Piscataway, NJ, USA, 2020; pp. 1–7.
30. Tizhoosh, H.R. Opposition-Based Learning: A New Scheme for Machine Intelligence. In Proceedings of the International Conference on Computational Intelligence for Modelling, Control and Automation and International Conference on Intelligent Agents, Web Technologies and Internet Commerce (CIMCA-IAWTIC'06), Vienna, Austria, 28–30 November 2005; IEEE: Piscataway, NJ, USA, 2005; pp. 6957–6961.
31. Roy, P.K.; Bhui, S. Multi-objective quasi-oppositional teaching learning based optimization for economic emission load dispatch problem. *Int. J. Electr. Power Energy Syst.* **2013**, *53*, 937–948. [CrossRef]
32. Sultana, S.; Roy, P.K. Multi-objective quasi-oppositional teaching learning based optimization for optimal location of distributed generator in radial distribution systems. *Int. J. Electr. Power Energy Syst.* **2014**, *63*, 534–545. [CrossRef]
33. Sharma, S.; Bhattacharjee, S.; Bhattacharya, A. Quasi-Oppositional Swine Influenza Model Based Optimization with Quarantine for optimal allocation of DG in radial distribution network. *Int. J. Electr. Power Energy Syst.* **2016**, *74*, 348–373. [CrossRef]
34. Yu, J.; Kim, C.-H.; Rhee, S.-B. Oppositional Jaya Algorithm With Distance-Adaptive Coefficient in Solving Directional Over Current Relays Coordination Problem. *IEEE Access* **2019**, *7*, 150729–150742. [CrossRef]
35. Abd Elaziz, M.; Mirjalili, S. A hyper-heuristic for improving the initial population of whale optimization algorithm. *Knowl. Based Syst.* **2019**, *172*, 42–63. [CrossRef]
36. Heidari, A.A.; Mirjalili, S.; Faris, H.; Aljarah, I.; Mafarja, M.; Chen, H. Harris hawks optimization: Algorithm and applications. *Future Gener. Comput. Syst.* **2019**, *97*, 849–872. [CrossRef]
37. Yadav, A. AEFA: Artificial electric field algorithm for global optimization. *Swarm Evol. Comput.* **2019**, *48*, 93–108.
38. Mehrabian, A.R.; Lucas, C. A novel numerical optimization algorithm inspired from weed colonization. *Ecol. Inform.* **2006**, *1*, 355–366. [CrossRef]

applied sciences

MDPI

Article

Economic Optimal Scheduling of Wind–Photovoltaic-Storage with Electric Vehicle Microgrid Based on Quantum Mayfly Algorithm

Ximu Liu, Mi Zhao *, Zihan Wei and Min Lu

College of Mechanical and Electrical Engineering, Shihezi University, Shihezi 832003, China
* Correspondence: zhaomi530@163.com

Abstract: The effectiveness of energy management systems is a great concern for wind–photovoltaic-storage electric vehicle systems, which coordinate operation optimization and flexible scheduling with the power grid. In order to save system operation cost and reduce the energy waste caused by wind and light abandonment, a time-sharing scheduling strategy based on the state of charge (SOC) and flexible equipment is proposed, and a quantum mayfly algorithm (QMA) is innovatively designed to implement the strategy. Firstly, a scheduling strategy is produced according to the SOC of the battery and electric vehicle (EV), as well as the output power of wind–photovoltaic generation. In addition, the minimum objective function of the comprehensive operation cost is established by considering the cost of each unit's operation and electricity market sale price. Secondly, QMA is creatively developed, including its optimization rule, whose performance evaluation is further carried out by comparisons with other typical bionics algorithms. The advantages of QMA in solving the low-power multivariable functions established in this paper are verified in the optimization results. Finally, using the empirical value of the power generation and loads collected in enterprise as the initial data, the mayfly algorithm (MA) and QMA are executed in MATLAB to solve the objective function. The scheduling results show that the time-sharing scheduling strategy can reduce the system's cost by 60%, and the method decreases energy waste compared with ordinary scheduling methods, especially when using QMA to solve the function.

Keywords: microgrid; economic scheduling; clean energy; quantum mayfly algorithm (QMA)

Citation: Liu, X.; Zhao, M.; Wei, Z.; Lu, M. Economic Optimal Scheduling of Wind–Photovoltaic-Storage for Electric Vehicle Microgrid Based on Quantum Mayfly Algorithm. *Appl. Sci.* **2022**, *12*, 8778. https://doi.org/10.3390/app12178778

Academic Editors: S. M. Muyeen and Mohamed Benbouzid

Received: 7 August 2022
Accepted: 28 August 2022
Published: 31 August 2022

1. Introduction

With the rapid development of the national economy, demand for fossil fuels is increasing. However, the output of traditional energy is limited, and the utilization of renewable energy has become the general trend. As new energy sources, wind and light are widely distributed and can be permanently used. They are characterized by low cost and strong environmental protection. Nevertheless, there are considerable random fluctuations of output power due to natural conditions, which affect the security and stability of the power system after grid connection [1–3]. Therefore, designing a stable and low-cost scheduling strategy based on the structure of the wind–photovoltaic-storage electric vehicle complementary power generation system has become a key issue. By coordinating wind–photovoltaic and flexible devices, we can ensure the reliability of the system, improve the economy, and increase environmental friendliness [4,5].

In the process of modeling, parameter uncertainty becomes a significant problem that threatens system security with the increasing scale of the microgrid. In order to solve this problem, robustness becomes one of the parameters worthy of study [6,7]. The improvement of robustness can make the system maintain its balance when it is disturbed, thereby reducing the loss caused by disturbances [8]. Meanwhile, some other power consumption units are introduced, such as electric vehicles (EVs) and fuel cells, on the basis of wind–photovoltaic-storage microgrid architecture, which can ensure the system maintains a stable operation by interacting power with other components when it is impacted [9,10]. In terms of the objective function, some microgrid systems, including fossil fuels, need to be considered regarding carbon emissions to reduce the environmental pollution caused by them [11]. Therefore, multi-objective functions combined with economic objectives and carbon emissions are established in recent studies. Furthermore, in order to ensure the normal operation of the microgrid system, it is necessary to ensure power conservation is met, and power flow constraints, including power flow calculation, should be added to the scheduling study [12,13]. Additionally, equipment parameters are important constraints to enable the system to operate under the superior performance of the equipment, thereby ensuring equipment safety and power quality.

In order to realize the scheduling strategy and accurately find the optimal value of the objective function, a variety of high-performance intelligent algorithm schemes based on biological habits are proposed, and verified in specific cases [14–16]. Bionics algorithms, such as the flying sparrow search algorithm, flower pollination algorithm, and other heuristic algorithms, are proposed and applied to practical problems [17–19]. Among them, as a new bionics algorithm, the mayfly algorithm (MA) has better optimization performance than the traditional intelligent algorithms. The integration of the advantages of various algorithms has significantly improved the optimization process in terms of speed and accuracy [20]. However, when it comes to high-dimensional complex problems, it is still difficult to jump out of local optimal regions simply by relying on its own mechanism, and the convergence accuracy of the algorithm is not high enough because the step size of the mayfly is too short when it moves. Based on the above shortcomings, logistic mapping is used to improve the optimization ability of mayfly individuals [21,22], or the learning factor is changed to improve the degree of self-adaptation [23].

In order to describe the research status of microgrid scheduling and its characteristics more clearly, the studies conducted for our paper, and other related studies, are summarized in Table 1, which is represented as follows:

Table 1. Comparison of relevant studies.

Number	EV	Robustness	Algorithm	Pollution	Market	Remark
[6]		√		√		ANN-based scheduling control approaches
[7]		√			√	Proposes a robust model predictive control approach
[9]	√	√		√		Addresses the uncertainty of PV output and EV charging
[10]					√	Solves the sub-problems with fitted Q-iteration
[11]			√	√	√	Uses improved algorithm to mine magnesium energy
[12]					√	Introduces a non-cooperative framework
[17]			√			Improvements and comparisons of algorithms
[18]	√		√		√	Uses ASAPSO algorithm in multi-objective optimization
[19]			√		√	Plans two-stage form of multi-energy supply optimization
This paper	√		√		√	Improves MA algorithm and designs a scheduling model

Through the above table, it can be seen that the robustness and stability of the system are studied by some scholars. In these studies, robust model predictive control (RMPC) and other methods are used to deal with the uncertainties of renewable energy. In the microgrid system containing fossil energy, reducing carbon emissions is also an important issue that needs to be solved to improve the environmental protections of the system. More studies focus on the microgrid system containing EVs, namely the EVs charging and discharging states, as well as the stability of the EVs themselves.However, it is also worth studying how to interact with the microgrid while ensuring its stable operation. Meanwhile, some studies are aimed at improved algorithms, which have been greatly modified in optimization rate, convergence, and escape from local deadlock; however, there is still room for further improvement. Therefore, the problems that are concentrated on in this paper are represented as follow:

(1) The EV is taken as one of the dispatching objects, and its interaction state with the microgrid is judged according to its SOC. The microgrid system is able to make full use of power in this measure.

(2) The QMA algorithm raised in this paper is compared with the moth–flame optimization (MFO), grey wolf optimize (GWO), particle swarm optimization (PSO), whale optimization algorithm (WOA), sine cosine algorithm (SCA), and MA to verify its advantages, and is applied to the solution of the objective function.

The rest of the paper is organized as follows: The structure model of the microgrid is established in Section 2. Meanwhile, the scheduling strategy of each unit, as well as the objective functions with constraint conditions of economic dispatching regarding the microgrid, are established. In Section 3, the operation rules of the QMA are analyzed, and the comparison between various intelligent algorithms is realized. The algorithms are combined with the objective function in Section 4, and the economic scheduling of the microgrid in a specific area is carried out according to the actual situation. Section 5 summarizes the content and contributions of this paper, and puts forward prospects for future studies.

2. Microgrid Structure and Dispatching Strategy

2.1. Microgrid Structure

The model of the wind–photovoltaic-storage electric vehicle microgrid system is shown in Figure 1. Power output can be achieved by wind turbines and photovoltaic generations, while the load can only consume power. As flexible equipment, the consumption and production of electric energy can be realized by EVs and batteries. Meanwhile, energy interaction can be implemented by the accession of the microgrid to the main grid .

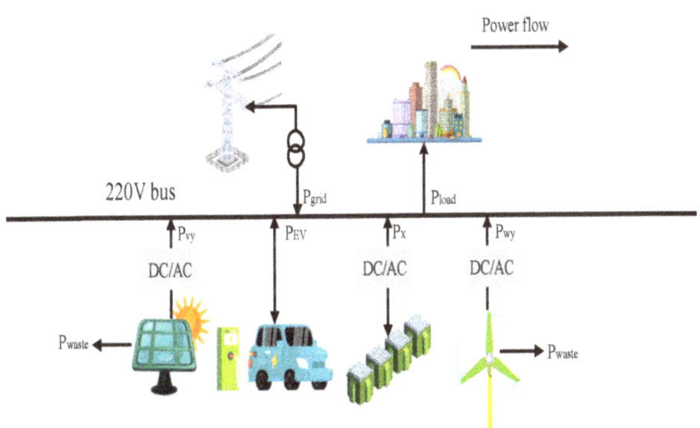

Figure 1. The model of wind–solar-storage electric vehicle microgrid system.

2.2. Scheduling Strategy Research

In this paper, in order to minimize the overall cost by dispatching each generation unit, the following requirements should be met in the scheduling strategy to ensure power supply reliability:

(1) Maximize the utilization of wind and photovoltaic power generation, and maximize their output through reasonable scheduling;
(2) Minimize the redundant power generated by wind and photovoltaic units to avoid energy waste;
(3) The difference between wind–photovoltaic generation and the power consumption of the load depends on the energy interaction with batteries, EVs, and the main grid, and whether this can meet power balance constraints.The comprehensive cost can be further reduced to achieve economic optimization by selling electricity to the main grid and EVs.

In order to achieve the above requirements, the dispatching strategy aiming at the system economy is designed as follow: The output of wind power and photovoltaic generations are coordinated, and participate in the dispatching according to the state of the battery and EV on charge. Finally the power balance is realized through the interaction process with the main grid.

a. Wind and photovoltaic generation

The empirical output value of wind and photovoltaic generation is taken as the actual output power, and the scheduling is carried out after increasing the cost of wind and solar abandoning in order to reduce energy waste. The output power of wind and photovoltaic generation after dispatching is compared with the empirical value. If the output power is lower than the empirical value, the output power after dispatching can be used; otherwise, the empirical value has to be maintained.

b. Battery generation

Since photovoltaic and wind generation is affected by daily environmental changes, the charging and discharging process of batteries can meet the demand. According to power load trends within 24 h, power supply states are divided into peak, normal, and valley periods. Charging and discharging strategies are adopted based on the current period and the battery state of charge (SOC). The specific process is shown in Figure 2.

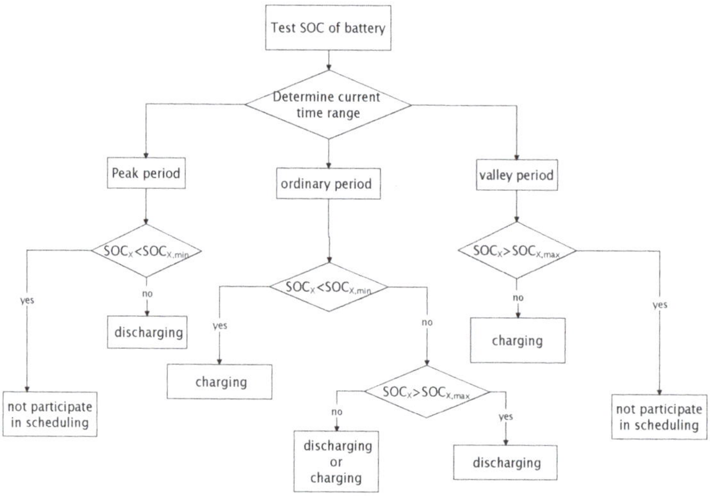

Figure 2. The scheduling strategy of battery generation.

According to Figure 2, charging and discharging can be carried out only if the battery SOC is within a certain range. When the charging and discharging standard is reached by the battery, the interactive state is judged according to its SOC value and the time period. Thereby, the charging and discharging power is calculated. The specific calculation method is provided as follow:

$$P_{bess}(t) = SOC_x(t)C_x \tag{1}$$

where the charging and discharging power of battery at time T is described as $P_{bess}(t)$. The SOC of the battery at time T is shown as $SOC_x(t)$, which is the scheduling target of the battery. The rated capacity of the battery collected from empirical data is defined as C_x.

c. Electric vehicle

The output of an EV can be judged according to its SOC. If the SOC can maintain the normal operation of the vehicle, the power is outputted to the system; otherwise, the system's power is consumed. The flow is expressed in Figure 3.

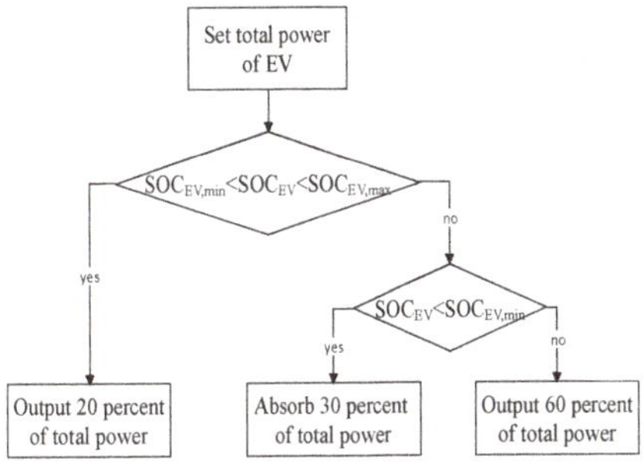

Figure 3. The scheduling strategy of EV.

Similar to battery scheduling, the output state of EVs can be judged according to their SOC. The output of the SOC after scheduling, according to the idea shown in Figure 3, can be used for further calculations, and the specific expression is represented as follows:

$$P_{ev}(t) = SOC_e(t)C_e \tag{2}$$

where interactive power between EVs and the microgrid at time T is given as $P_{ev}(t)$. The SOC of the EV at time T is shown as $SOC_e(t)$, and the fixed capacity of EV is expressed as C_e.

2.3. *Expression Construction*

2.3.1. Objective Function

In this paper, objective functions are established for the economy of the wind–photovoltaic-storage electric vehicle microgrid system under a grid-connected operation. Consider the operation, maintenance, and investment cost of each unit in a system, and the cost of grid sales and purchases. Meanwhile, taking into account the operating costs of EVs and the cost policies issued by the government for new energy, the optimization goal can be achieved through relevant scheduling. The expression is as follows:

$$Minimize\ F(x) = F_{wv} + F_x - F_G - F_{EV} + F_{BT} \tag{3}$$

where the operation and maintenance costs of wind and photovoltaic generation are represented as F_{wv}. The power generation and investment costs of the battery are determined as F_x. The power grid sales and purchase costs are simplified as F_G. The operating costs of EVs and policy expenses related to new energy power generation are given as F_{EV} and F_{BT}. The decision variable of the objective function is the output power of each unit, which is the real number. Moreover, the specific coupling relationships between decision variables and objective functions (3) are shown in subsections a–e.

a. Wind and photovoltaic generation

The equipment loss, manual repair, and operating reserve storage costs are included in F_{wv}, which can be expressed as follows:

$$F_{wv} = \sum_{t=1}^{T} C_w P_{wf}(t) + \sum_{t=1}^{T} C_v P_{vf}(t) + F_{RE} \tag{4}$$

where the operation and maintenance cost coefficients of wind and photovoltaic generation are defined as C_w and C_v, respectively. These coefficients are based on the ratio of the total operation and investment costs of the equipment to the total annual generated power within a year obtained in enterprise. The output power of wind and photovoltaic power generation is computed as P_{wf} and P_{vf}, respectively. The operating reserve storage cost is simplified as F_{RE}, which can be expressed as follows:

$$F_{RE} = C_{RE}[P_{vf}(t) + P_{wf}(t)] \tag{5}$$

where the operating reserve storage cost coefficients of wind and photovoltaic generation are given as C_{RE}.

b. Battery generation

In Equation (6), F_x consists two parts, which are the acquisition and maintenance costs of batteries. The acquisition cost is converted into each work process, and the specific expression is calculated as follows:

$$F_x = \frac{F_{cap}C}{P_{bess}T_n}|\sum_{t=1}^{T} P_{bess}(t)| + \frac{K_o P_{bess,max}}{365} + K_M \frac{1}{\Delta t} P_{bess}(t)\Delta t \tag{6}$$

where the acquisition cost and depreciation coefficient are defined as F_{cap} and C, respectively. The annual operating cost coefficient and hours are represented as Ko and T_n, respectively. P_{bess} and $P_{bess,max}$ are battery-rated power and maximum charge and discharge power, respectively. Meanwhile, the maintenance cost coefficient is described as K_M, and the duration from the beginning to the end of charging and discharging the battery is shown as T.

c. Interaction with main grid

The microgrid system can interact with the main grid when the power is vacant or redundant. When the interactive power is positive, the power is purchased from the main grid; otherwise, the interaction cost F_G is the sum of the two. The equation is simplified as follows:

$$F_G = \sum_{t=1}^{24} P_{grid}(t) C_{price,grid} \tag{7}$$

where $P_{grid}(t)$ refers to interactive power with the main grid, and the time-sharing sale and purchase price of the main grid is expressed as $C_{price,grid}$.

d. Policy of government

Subsidies are paid for wind and photovoltaic generation, which encourage companies to adopt new sources of electricity. Meanwhile, fees are charged for energy waste to reduce the energy loss caused by wind and solar discarding. The formula is computed as follows:

$$F_{BT} = \sum_{t=1}^{T} C_q[P_{wy}(t) + P_{vy}(t) - P_{wy}(t) - P_{vy}(t)] - S_{bt} \tag{8}$$

where the wind and solar discarding coefficient is given as C_q and S_{bt}, which represents the government subsidy. The actual utilization power of wind and photovoltaic generation is determined as $P_{wy}(t)$ and $P_{vy}(t)$, respectively.

e. Electric vehicles

As a client device, the cost of selling and purchasing power to the microgrid is mainly included in the cost of EVs. The specific expression is given as follows:

$$F_{EV} = \sum_{t=1}^{24} P_{EV}(t) C_{price,EV} \tag{9}$$

where the electricity purchase price of an EV is defined as $C_{price,EV}$.

2.3.2. Constraint Condition

In order to ensure the safe and stable operation of the grid system, the power balance should be regarded as the basic constraint condition, and the output power and SOC of flexible devices should be limited to ensure the normal operation of equipment.

a. Power balance

In any period of time, all output, load, flexible device charging and discharging, and interaction powers between the microgrid system and the main grid should be kept in balance, and the constraint equation is given as follows:

$$P_{wy}(t) + P_{vy}(t) + P_{grid}(t) + P_x(t) + P_{EV}(t) = P_{load}(t) \tag{10}$$

where P_x is the charging and discharging power of the battery, and the total load is expressed as P_{load}.

b. Wind and photovoltaic generation

Due to the limitations of the equipment parameters, the output power of wind and photovoltaic generation is within a certain range. Meanwhile, in order to ensure the normal operation of the system when the power required surges, it is necessary to add operating reserves. In detail, 30% of power generated by wind and photovoltaic power generation is stored, and the rest is the maximum power supplied to the system. Furthermore, the change of power in each time period should also be constrained in a certain range. Therefore, the relevant constraints are set as follows:

$$\begin{cases} 0 \le P_{wy}(t) \le 0.7P_{wf}(t) \\ 0 \le P_{vy}(t) \le 0.7P_{vf}(t) \\ R_{w-down}\delta t \le \delta P_w \le R_{w-up}\delta t \\ R_{v-down}\delta t \le \delta P_v \le R_{v-up}\delta t \end{cases} \tag{11}$$

where the adjustable power amplitude of wind and photovoltaic generation in a certain period is represented as δP_w and δP_v, respectively. The downward climbing rate of wind and photovoltaic generation during adjustment is expressed as R_{w-down} and R_{v-down}, respectively. Meanwhile, R_{w-up} and R_{v-up} give the upward climbing rate for wind and photovoltaic generation during adjustment, respectively. The adjustment time is represented by δt.

c. Flexible generation

Batteries are not only constrained by power, their safety performance and service life are affected by their SOC , and the operation of an EV is also affected by its SOC. Specific conditions are simplified as follows:

$$\begin{cases} SOC_{emin} \le SOC_e(t) \le SOC_{emax} \\ SOC_{xmin} \le SOC_x(t) \le SOC_{xmax} \\ 0 \le P_{bess,t}(t) \le P_{bess,max} \end{cases} \tag{12}$$

SOC_{emin}, SOC_{xmin}, SOC_{emax}, and SOC_{xmax} are the upper and lower limits of their SOC.

In addition, to ensure the safe and stable operation of batteries, the charging and discharging rates should be controlled within a certain range. The specific constraints are shown as follows:

$$\delta SOC_x \le 0.2 \tag{13}$$

where the adjustable range of a battery's SOC in each period is expressed as δSOC_x.

3. Improved Mayfly Algorithm

3.1. Traditional Mayfly Algorithm

The MA is a bionics algorithm derived from the social behavior of the mayfly. Inspired by the movement mode and reproduction process of female and male populations, the optimal and suboptimal individuals in each population are selected. Meanwhile, the optimal offspring generation is obtained through mating between the optimal male and female individuals, and the suboptimal offspring generation is obtained in the same way. The direction of movement of each mayfly is influenced by the dynamics of individual and collective optimal positions, and female mayflies target male mayflies towards their positions.

a. The movement of male mayfly

The flight mode of male mayflies is similar to the movement mode of birds in a particle swarm algorithm, and the direction and distance of male mayflies are adjusted according to their own flight experiences and that of individuals around them. The specific method is shown as follows:

$$x_i^{n+1} = x_i^n + v_i^{n+1} \tag{14}$$

where x_i^n and v_i^n are the current position and speed, respectively, of the male mayfly i on the nth search. The values are calculated as follows:

$$v_{ij}^{n+1} = v_{ij}^n + \alpha_1 e^{-\beta l_p^2}(p_{best_i^j} - X_{ij}^n) + \alpha_2 e^{-\beta l_g^2}(g_{best_i^j} - X_{ij}^n) \tag{15}$$

Because male mayflies perform a dance on the surface of water to attract females, the position of the male mayflies is constantly changing, meaning they do not build a high speed. v_{ij}^n is the speed of the nth search of the mayfly i at j dimension, and x_{ij}^n is the position at that time. α_1 and α_2 are estimated from the positive attraction coefficients of social interaction, and β is the visibility coefficient of the mayfly. Meanwhile, the optimal locations of the individual and collective mayflies are expressed as $p_{best_i^j}$ and $g_{best_i^j}$, respectively. Additionally, the distances from current position to $p_{best_i^j}$ and $g_{best_i^j}$ are defined as l_p and l_g, respectively, and are calculated as follows:

$$\|x_i - X_i\| = \sqrt[2]{\sum_{j=1}^{n}(x_{ij} - X_{ij})^2} \tag{16}$$

For the best mayfly in the population, a fixed dance pattern is needed to be maintained. Meanwhile, a random element is introduced to ensure that the speed is constantly changing. The calculation is described in this case as follows:

$$v_{ij}^{n+1} = v_{ij}^n + d \times r \tag{17}$$

where the dance coefficient is simplified as d, and r is the random natural number within $[-1, 1]$.

b. The movement of female mayfly

The movement of the female mayfly depends on the attraction of the male mayfly, and their position renewal depends on the increase of speed, which can be expressed as follows:

$$y_i^{n+1} = y_i^n + v_i^{n+1} \tag{18}$$

The speed update is a certain process, which means that in order to ensure the quality of offspring, the best female needs to be attracted by the best male, the second-best female by the second-best male, and so on. The speed update expression is given as follows:

$$v_{ij}^{n+1} = \begin{cases} v_{ij}^n + \alpha_2 e^{-\beta l_f^2}(x_{ij}^n - y_{ij}^n) \\ v_{ij}^n + g \times r \end{cases} \tag{19}$$

where the position of the female mayfly is expressed as y_{ij}^n, the random walk coefficient of the female mayfly is represented as g, and l_f is determined by the distance between the male and female mayflies.

c. The mating of male and female mayflies

In the process of parent mating, the optimal and suboptimal individuals in the male and female populations should be selected for mating and reproduction based on their fitness functions. The results of interbreeding, which produces the optimal and suboptimal offspring, are calculated as follows:

$$\begin{cases} offspring1 = L \times m + (1-L) \times f_m \\ offspring2 = L \times f_m + (1-L) \times m \end{cases} \quad (20)$$

where the male and female in the parent generation are represented as m and f_m, respectively, and L is a random natural number within a specific range.

3.2. Ma with Quantum Idea

The traditional MA can find the optimal value in a single-peak function accurately by taking advantage of the characteristics in mayfly reproduction. However, with regard to the large population and complicated process, the search speed is slow and the convergence is not fine. Meanwhile, it is easy to fall into local deadlock when dealing with multi-peak functions. Therefore, the quantum idea is introduced on the basis of a traditional MA, thereby forming the QMA. Because the position and velocity of the mayfly cannot be determined simultaneously in quantum space, the wave function is used to represent the position of the mayfly, and the Monte Carlo method is used to solve the problem. The particle update expression is shown as follows:

$$\begin{cases} m_{best^n} = \frac{1}{N}\sum_{i=1}^{a} P_{best_i^n}(i = 1 \ldots n) \\ P_i^t = \gamma \times P_{best_i^n} + (1-\gamma) \times g_{best}^n \\ X_i^{n+1} = P_i^n \pm \varepsilon |m_{best^n} - x_i^n| \log \frac{1}{a} \end{cases} \quad (21)$$

where the numbers of individuals and iterations are represented as N and n, respectively. m_{best^n} is the average historical optimal position of the male mayfly, and P_i^n is obtained from the updated position of the ith male mayfly in the nth iteration. r and a are uniformly distributed values within (0,1), and c is the final random motion parameter.

The execution steps of the QMA are given as follows:

Step 1: Initialize the positions of the female and male mayflies in the space.

Step 2: Calculate the average optimal location m_{best} of the male mayflies according to the first equation of (21).

Step 3: Calculate the fitness value and sorting according to Formula (3), and compare with the previous iterative value. If the current function value is less than the previous iteration, the current mayfly position is updated for the individual optimal position, otherwise it keeps the previous iteration. Hence, the optimal male individual p_{best} and collective locations g_{best} are obtained.

Step 4: Calculate the new positions of the female and male mayflies according to Formula (18) and the second equation of (21), respectively, and mate in sequence.

Step 5: Calculate the fitness function and update p_{best} and g_{best}.

Step 6: Repeat Step 2 to 5 until the stop condition is met.

The QMA flow chart is shown in Figure 4.

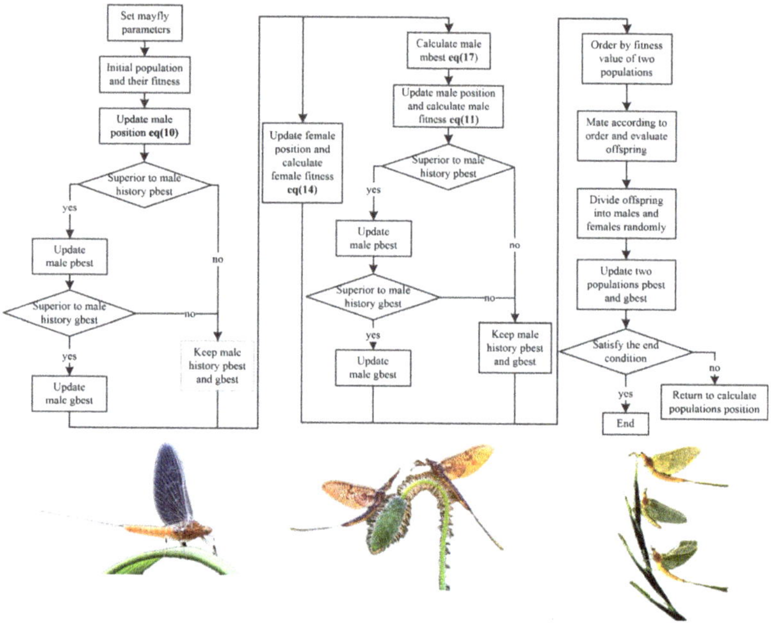

Figure 4. The flow chart of QMA.

3.3. Performance Analysis of Qma

In order to verify the superiority of the QMA, the single-peak and multi-peak functions are tested by MFO, GWO, PSO, WOA, SCA, MA, and QMA in this section. We combine the biological habits and existing literature on each algorithm, and the initialization parameters and test functions are shown in Tables 2 and 3, respectively. Meanwhile, to ensure the effectiveness of the comparison results, the number of biological populations in all bionics algorithms is set as 10. The function expression is represented by *Fun*, and the expression dimension is expressed as *Dim*. L_b and U_b are the upper and lower boundaries of the variables, respectively.

Table 2. Algorithm parameters.

Type	Parameters	Value	Parameters	Value
QMA/MA	Mutation rate r_m	0.01	Personal learning coefficient l_n	1
	Global learning coefficient l_g	1.5	Inertia weight w	0.8
	Distance sight coefficient d	2	Nuptial dance coefficient n	5
	Random flight coefficient r	1	Damping ratio d_a	0.99
WOA	Inertia weight a	20→0	Inertia weight a_1	−10→−20
MFO	Maximum weight a_{max}	−1	Minimum weight a_{min}	−2
GWO	Inertia weight a_{max}	20	Minimum weight a_{min}	0
SCA	Inertia weight a_{max}	2	Minimum weight a_{min}	0
PSO	Maximum velocity v_{max}	6	Minimum velocity v_{min}	−6
	Maximum weight w_{max}	0.9	Minimum weight w_{min}	0.2
	Inertia weight c	2	Inertia weight c_1	2

Table 3. Test functions.

Number	Function	Dim	L_b	U_b	Iteration
F1	$f(x) = \sum_{i=1}^{n} x_i^2$	30	−100	100	1000
F2	$f(x) = An + \sum_{i=1}^{n} [x_i^2 - A\cos(2\pi x_i)]$	30	−5.2	5.2	1000

Sphere function is represented as F1, which is a typical single-peak function. The optimization time and local search ability of the algorithm can be effectively verified in this type of function. The Rastrigrin function is simplified as F2, and the validity of breaking away from local deadlock in the algorithm can be proved in this multi-peak function. The performances of different optimization algorithms in solving these two functions are expressed in Table 4. The iteration curves of these functions run once are shown in Figures 5 and 6.

Table 4. Comparison results in different algorithms.

Number	Type	MFO	WOA	SCA	GWO	PSO	MA	QMA
F1	Average	5693.2166	1382.2805	4.6177274	0.2631747	0.01161598	5.10×10^{-8}	7.51×10^{-9}
	Var	44,354,655.8	1,791,398.94	309.74423	0.17328	0.00112851	1.92×10^{-15}	3.33×10^{-16}
	STD	6659.92911	1338.43152	17.599552	0.4162691	0.03359328	4.38×10^{-8}	1.83×10^{-8}
F2	Average	197.08007	109.26403	31.40007	37.90619	132.70364	33.91202	20.92731
	Var	2051.98830	6157.62175	772.52064	171.00581	1369.34713	114.93251	39.28933
	STD	45.29888	78.47052	27.79426	13.07692	37.00469	10.72066	6.26812

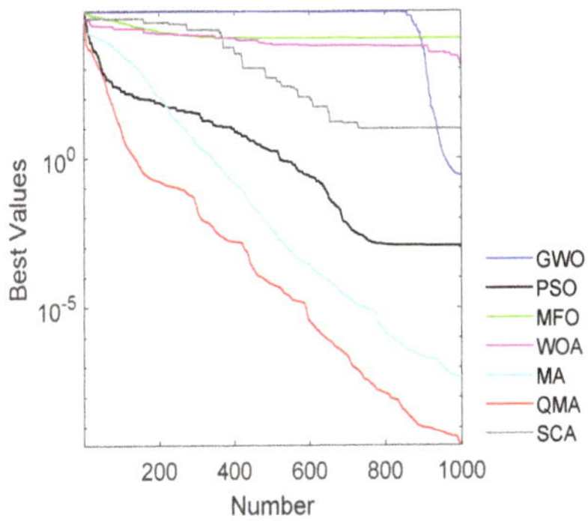

Figure 5. The optimization of algorithms using Sphere function.

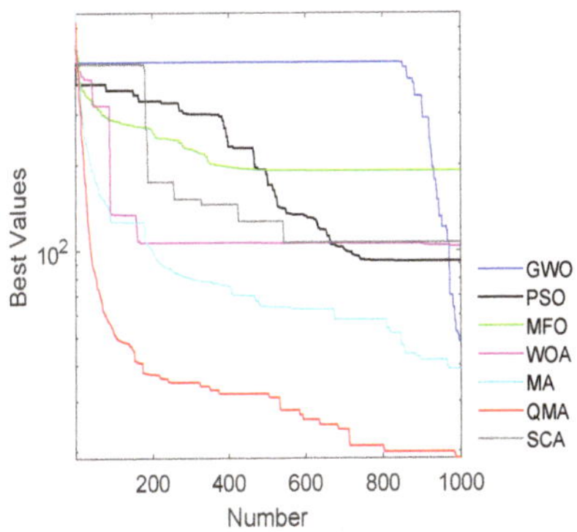

Figure 6. The optimization of algorithms using Rastrigrin function.

As can be seen from the Figures and Table, the average value, variance, and standard deviation of the different algorithms after 30 iterations in solving functions are different. The optimal value can be found quickly in the WOA, MFO, and GWO algorithms once they get rid of local deadlock when solving multi-variable functions. However, it is easy for them to fall into a local optimum, which is greatly affected by the number of iterations and randomness. Hence, it is not suitable for these algorithms to solve such functions. Compared with the SCA, PSO, and MA algorithms, although the convergence speed is not fast as the above algorithms, the rate is stable, and it is not easy to get into local deadlock. They have good performance in solving multi-variable lower-power functions, and MA has the most stable optimization rate among them. On the basis of MA, the stability of optimization is improved further in the QMA, and the optimal solution of the objective function can be found better in it. In the function where the optimal value is zero, the optimal value found by the QMA in the optimization process is lower than other algorithms, and the performance is more stable with little variance. In the scheduling problem of the microgrid, many variables are obtained in an economic objective function, and the stability of optimization is needed to ensure this. Hence, it is suitable for the QMA to find the optimal value of the economic scheduling function based on the microgrid established in this paper.

4. Simulation Analysis of Microgrid

Entering into cooperation with an enterprise, a typical winter day in the region where the enterprise is located is taken as an example to make economic scheduling. The actual data, such as wind–photovoltaic power and residential electricity consumption obtained from enterprise, are input as empirical values. Based on the economic scheduling objective function of the wind–photovoltaic-storage electric vehicle microgrid established in Section 2, the QMA and MA algorithms are used to solve it. The algorithms are invoked, combining with the microgrid economic objective function, and the number of the population in the algorithms is set as 100 after several simulations. In this condition, the objective function the algoriths can commendably solve the problems regarding optimization speed and convergence. Hence, the output power of each unit within 24 h is obtained. The output empirical values of wind and photovoltaic generation collected from enterprise are expressed in Figure 7, and the empirical values of residential electricity consumption are

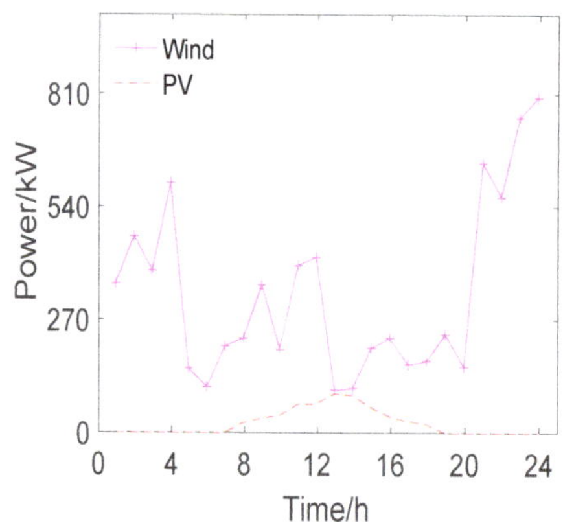

Figure 7. Empirical values of wind and photovoltaic generation.

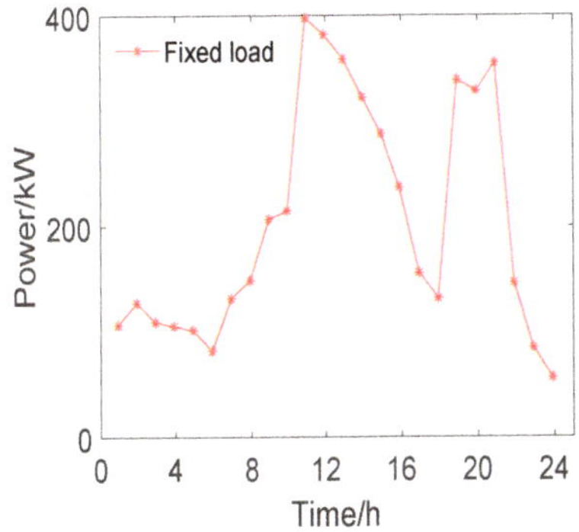

Figure 8. Empirical values of fixed load.

Under the load conditions, the optimization algorithms are used in this paper to find the minimum output cost under the condition of maintaining the security and stability of the system. Combined with empirical data obtained from the enterprise, the status of various components in the microgrid and the equivalence coefficient are shown in Tables 5 and 6. The related parameters of the time-of-use price are shown in Table 7.

Table 5. Status of various components in the microgrid.

Type	Maximum Output Power/kW	Maximum Ramp/kW	(Min, Max) SOC
Wind generation	810	50	\
Photovoltaic generation	100	50	\
Battery	1000	200	0.3,0.7
Charging power of EV	120	\	0.3,0.7
Main grid transmission	1000	\	\

Table 6. Equivalence coefficient.

Parameters	Value
Photovoltaic generation operation and maintenance coefficient (RMB/kW)	0.0096
Wind generation operation and maintenance coefficient (RMB/kW)	0.0296
Battery annual operating cost factor (RMB/kW)	50
Battery maintenance cost factor (RMB/kW)	0.813
Battery depreciation factor (RMB/kW)	0.005
Battery annual operating hours (hours)	50
Battery investment cost (RMB/kW)	5000
Wind and light abandoning coefficient (RMB/kW)	0.025
Government subsidy (RMB/kW)	50
Operating reserve storage cost coefficient (RMB/kW)	0.002

Table 7. Market parameters.

Selling Object	Time	Price (RMB)
EV	0:00–24:00	0.5
Load	0:00–7:00	0.49
	7:00–10:00	0.83
	10:00–14:00	1.1
	14:00–19:00	0.83
	19:00–22:00	1.1
	22:00–24:00	0.49

When MA and QMA are used to solve the objective function, the SOC of the battery and EV are shown in Figures 9 and 10, respectively . The recharging of the battery is represented as an increase in SOC; otherwise, the battery delivers power to the grid. When the SOC of the battery is between 0.3 and 0.7, it can maintain its operation in the best condition. Taking Figure 10 as an example, when QMA is used to solve the system objective function, the battery is charged in three time periods: 1:00–2:00, 15:00–16:00, and 23:00–24:00. Meanwhile, discharge is performed in four time periods: 10:00–12:00, 14:00–15:00, 18:00–19:00, and 20:00–21:00. The rest of time, the battery remains neither charged nor discharged. Furthermore, if there is more than one EV in the area, their SOC is not continuous with great fluctuations, and they can operate normally when the SOC is within the range of 0.3 to 0.7.

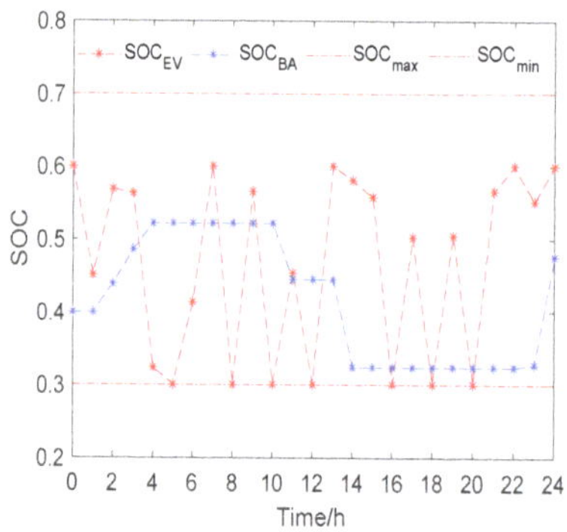

Figure 9. The battery and EV SOC under the objective function solving by MA.

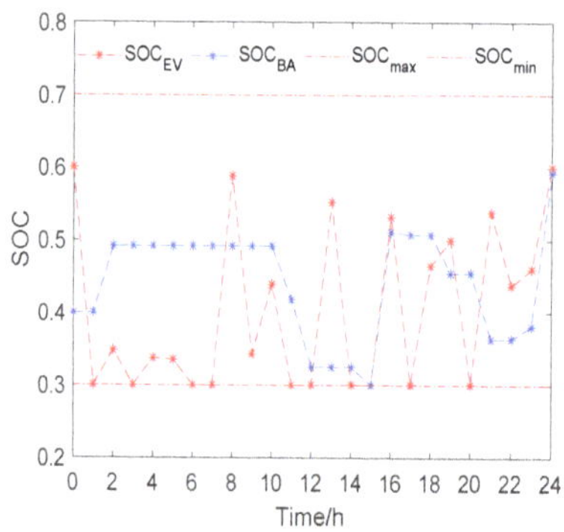

Figure 10. The battery and EV SOC under the objective function solving by QMA.

The output of each unit when MA and QMA solve the objective function is represented in Figures 11 and 12, respectively. The load power curve is the absolute value of consumed power.

Abundant power is generated by wind turbines within 0:00 to 4:00, and the microgrid consumes redundant power by interacting with the main grid. From 4:00 to 11:00, the output power of wind–photovoltaic generation and EVs supplies power to load; meanwhile, the redundant power is consumed by the main grid. From 11:00 to 16:00 is the peak period of electricity consumption. During this time, almost all power generation is used to maintain the power required by load. The load demand gradually decreases from 16:00 to 18:00 and reaches the peak again from 18:00 to 21:00. In this period, wind generation and EVs are in a power output state, while photovoltaic generation occasionally outputs power, and the interaction with the main grid absorbs or transfers power to the main grid as the load requires. Finally, load demand drops from 21:00 to 24:00, and wind generation and EVs mainly send out power and feedback to the main grid.

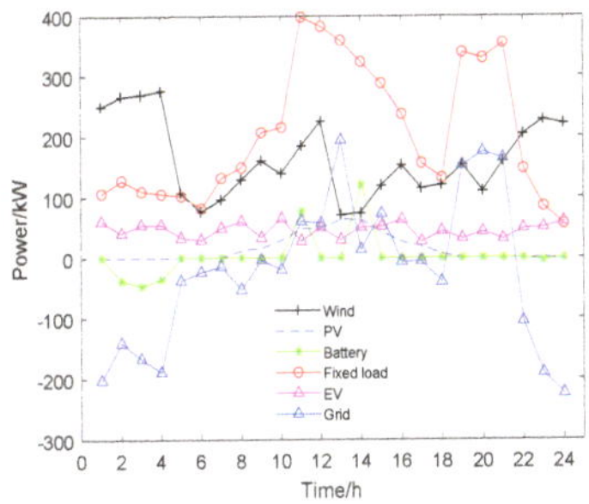

Figure 11. The output of each unit under the objective function solving by MA.

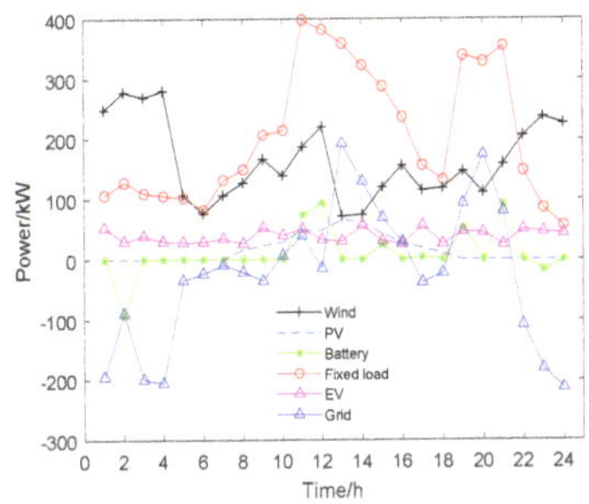

Figure 12. The output of each unit under the objective function solving by QMA.

When the output power is at 10% of its maximum capacity it is stored by the battery, and the wind–photovoltaic generation produces power according to the empirical value. The operation cost of the microgrid is shown in Figure 13; furthermore, the operation cost of the microgrid obtained by solving the objective function using the scheduling strategy in Section 2 is shown in Figure 14.

The results of the value and standard deviation of MA and QMA after 100 and 500 iterations are shown in Table 8. In terms of these results, the operating costs under the empirical value are higher than the costs after participating in scheduling according to the strategy. Moreover, the optimization rate of QMA is higher than that of MA, and the optimal solution found by the improved algorithm after 500 iterations is obviously better than the solution obtained before the improvement.

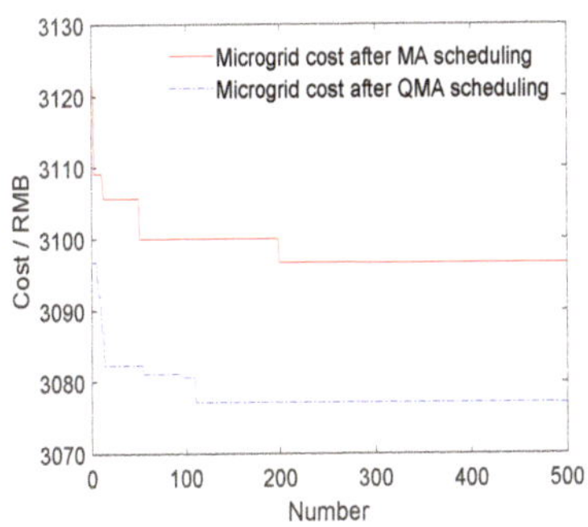

Figure 13. Microgrid operation cost utilizing regular scheduling mode.

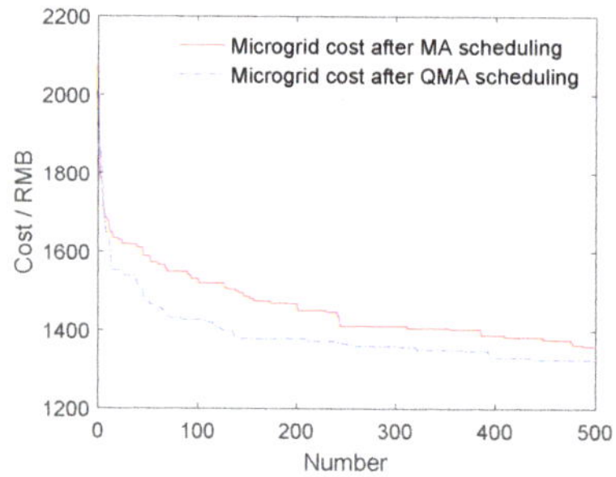

Figure 14. Microgrid operation cost after time-sharing scheduling.

Table 8. The cost of microgrid.

Type Iterations	Time-Sharing Scheduling				Normal Scheduling			
	QMA		MA		QMA		MA	
	100	500	100	500	100	500	100	500
Mean(RMB)	1460.53	1310.10	1498.17	1328.32	3078.50	3073.22	3098.10	3093.96
Var	833.71	718.22	3080.84	761.05	23.83	9.13	6.14	5.76
STD	28.87	26.79	55.50	27.58	4.88	3.02	2.47	2.40

In summary, the scheduling strategy provided in this paper can effectively improve the economy of the microgrid system, and the improved QMA has a high level of superiority when solving the objective function.

5. Conclusions

According to the operation state of a wind–photovoltaic-storage electric vehicle microgrid connected to the grid in a certain area, a time-sharing scheduling strategy is designed in this paper; furthermore we established the minimum cost objective function according to economic requirements and constraints, such as wind and light abandonment. Meanwhile, a QMA algorithm is innovatively proposed and applied to the solution of the objective function. In summary, the main conclusions of this paper are expressed as follows:

(1) The SOC of the EV and battery is taken as a scheduling object, and it is scheduled in combination with the current period, thereby obtaining the interaction power between the flexible equipment and microgrid. Furthermore, the output power of wind and photovoltaic generation is limited within the empirical value and, accordingly, the energy waste caused by wind and light abandonment is reduced. This scheduling method not only reduces energy waste but also decreases the system cost by about 60%;

(2) On the basis of the MA, the quantum idea is introduced, and the QMA is proposed. Comparing the performance of the QMA with six other intelligent algorithms in typical functions optimization, the superiority of the QMA is clearly seen. Meanwhile, the QMA is applied to the objective function established in this paper. The cost of the system can be reduced by about 2%, which is considerable for the microgrid system.

In future research, new energy devices need to be taken into account, and the robustness of the system should also be paid attention to because of the increasing complexity of the system. Meanwhile, in order to make managers more intuitive, namely to understand the operation state of the system, the connection between information management and the physical layer should be strengthened. It is necessary to study the use of hierarchical scheduling, source and charge interactions, and other methods to meet the needs of different managers.

Author Contributions: X.L. wrote the manuscript, designed the scheduling strategy, drew the flow chart and table, and performed the simulation experiment and analysis. M.Z. supervised the study, helped obtain data, and checked the contents of the manuscript. Z.W. verified the simulation data and contributed to the writing and typesetting of the manuscript. M.L. provided practical suggestions and evaluated the feasibility of the application. All authors have read and agreed to the published version of the manuscript.

Funding: This work was Supported by the Natural Science Foundation of China under Grant No. 61563045.

Institutional Review Board Statement: Not applicable.

Informed Consent Statement: Not applicable.

Data Availability Statement: Not applicable.

Acknowledgments: Thanks to the electric power enterprises in Liaoning Province for providing data support for this research.

Conflicts of Interest: The authors declare no conflict of interest.

Acronyms

The following acronyms are used in this paper:

MA	Mayfly algorithm
QMA	Quantum mayfly algorithm
MFO	Moth–flame optimization
GWO	Grey wolf optimizer
WOA	Whale optimization algorithm
SCA	Sine cosine algorithm
PSO	Particle swarm optimization
EV	Electric vehicle
SOC	State of charge

References

1. Rui, T.; Li, G.; Wang, Q.; Hu, C.; Shen, W.; Xu, B. Hierarchical Optimization Method for Energy Scheduling of Multiple Microgrids. *Appl. Sci.* **2019**, *9*, 624. [CrossRef]
2. Wang, J.; Li, K.-J.; Javid, Z.; Sun, Y. Distributed optimal coordinated operation for distribution system with the integration of residential microgrids. *Appl. Sci.* **2019**, *9*, 2136. [CrossRef]
3. Harsh, P.; Das, D. Energy management in microgrid using incentive-based demand response and reconfigured network considering uncertainties in renewable energy sources. *Sustain. Energy Technol. Assess.* **2021**, *46*, 101225. [CrossRef]
4. Zhang, Z.; Wu, J.; Luo, Z.; Xu, J.; Jin, X.; Li, H. Optimal Scheduling for Independent AC/DC Hybrid Microgrid Considering Operation Characteristics of Energy Storage. *Power Syst. Autom.* **2018**, *42*, 118–125.
5. Hakimi, S.M.; Hasankhani, A.; Shafie-Khah, M.; Lotfi, M.; Catalão, J.P. Optimal sizing of renewable energy systems in a Microgrid considering electricity market interaction and reliability analysis. *Electr. Power Syst. Res.* **2022**, *203*, 107678. [CrossRef]
6. Abdolrasol, M.G.M.; Hannan, M.A.; Hussain, S.M.S.; Ustun, T.S.; Sarker, M.R.; Ker, P.J. Energy Management Scheduling for Microgrids in the Virtual Power Plant System Using Artificial Neural Networks. *Energies* **2021**, *14*, 6507. [CrossRef]
7. Carli, R.; Cavone, G.; Pippia, T.; De Schutter, B.; Dotoli, M. Robust Optimal Control for Demand Side Management of Multi-Carrier Microgrids. *IEEE Trans. Autom. Sci. Eng.* **2022**, *19*, 1338–1351. [CrossRef]
8. Melhem, F.Y.; Grunder, O.; Hammoudan, Z.; Moubayed, N. Energy Management in Electrical Smart Grid Environment Using Robust Optimization Algorithm. *IEEE Trans. Ind. Appl.* **2018**, *54*, 2714–2726. [CrossRef]
9. Shi, R.; Zhang, P.; Zhang, J.; Niu, L.; Han, X. Multidispatch for Microgrid including Renewable Energy and Electric Vehicles with Robust Optimization Algorithm. *Energies* **2020**, *13*, 2813. [CrossRef]
10. Mbuwirabc, B.; Spiessensab, F.; Deconinckac, G. Distributed optimization for scheduling energy flows in community microgrids. *Electr. Power Syst. Res.* **2020**, *187*, 106479–106490. [CrossRef]
11. De, M.; Das, G.; Mandal, K. An effective energy flow management in grid-connected solar–wind-microgrid system incorporating economic and environmental generation scheduling using a meta-dynamic approach-based multiobjective flower pollination algorithm. *Energy Rep.* **2021**, *7*, 2711–2726. [CrossRef]
12. Scarabaggio, P.; Carli, R.; Dotoli, M. Noncooperative Equilibrium Seeking in Distributed Energy Systems Under AC Power Flow Nonlinear Constraints. *IEEE Trans. Control Netw. Syst.* **2022**, 1–12. [CrossRef]

13. Yao, M.; Molzahn, D.K.; Mathieu, J.L. An optimal power-flow approach to improve power system voltage stability using demand response. *IEEE Trans. Control Netw. Syst.* **2019**, *6*, 1015–1025. [CrossRef]
14. Huang, C.; Zhao, K. Micro Grid Scheduling Optimization Based on Quantum Particle Swarm Optimization (PSO) Algorithm. *Power Syst. Clean Energy* **2016**, *2*, 72–76.
15. Yang, L.; Yang, X.; Wu, Y.; Wen, H.; Zhu, y. Research on optimized distributed generations locating based on modified cat swarm optimization. *Power Syst. Prot. Control* **2019**, *1*, 95–100.
16. Kong, X.; Xu, J.; Zhang, W. Ant Colony Algorithm of Multi-objective Optimization for Dynamic Grid Scheduling. *Metall. Min. Ind.* **2015**, *7*, 236–243.
17. Nguyen, T.; Ngo, T.; Dao, T.; Nguyen, T.T.T. Microgrid Operations Planning Based on Improving the Flying Sparrow Search Algorithm. *Symmetry* **2022**, *14*, 168. [CrossRef]
18. Mei, Y.; Li, B.; Wang, H.; Wang, X.; Negnevitsky, M. Multi-objective optimal scheduling of microgrid with electric vehicles. *Energy Rep.* **2022**, *8*, 4512–4524. [CrossRef]
19. Muhammad, H.; Wang, F.; Sultan, S.; Khan, Y.A.; Kalwar, B.A.; Fars, A. Two-stage intelligent planning with improved artificial bee colony algorithm for a microgrid by considering the uncertainty of renewable sources. *Energy Rep.* **2021**, *7*, 8912–8928.
20. Liu, C.; Li, Q.; Tian, X.; Wei, L.; Chi, Y.; Li, C. Multi-Objective Mayfly Optimization-Based Frequency Regulation for Power Grid With Wind Energy Penetration. *Front. Energy Res.* **2022**, *10*, 848966. [CrossRef]
21. Zervoudakis, K.; Tsafarakis, S. A mayfly optimization algorithm. *Comput. Ind. Eng.* **2020**, *145*, 106559. [CrossRef]
22. Jain, A.; Gupta, A. Review on Recent Developments in the Mayfly Algorithm. In Proceedings of the International Conference on Paradigms of Communication, Computing and Data Sciences, Singapore, 1 January 2022; pp. 347–357.
23. Dhawan, D.; Singh, R. Performance Evaluation of Nature Inspired Meta-Heuristic Algorithms using Rosenbrock, Rastrigin and Sphere Test Function for Optimization. *Int. J. Recent Technol. Eng.* **2019**, *8*, 1157–1163.

applied sciences

MDPI

Article

Smart Cooperative Energy Supply Strategy to Increase Reliability in Residential Stand-Alone Photovoltaic Systems

Carlos Roldán-Blay [1,*], Carlos Roldán-Porta [1], Eduardo Quiles [2] and Guillermo Escrivá-Escrivá [1]

[1] Institute for Energy Engineering, Universitat Politècnica de València, Camino de Vera, s/n, Building 8E, 46022 Valencia, Spain; croldan@die.upv.es (C.R.-P.); guieses@die.upv.es (G.E.-E.)
[2] Instituto de Automática e Informática Industrial, Universitat Politècnica de València, Camino de Vera, s/n, 46022 Valencia, Spain; equiles@isa.upv.es
* Correspondence: carrolbl@die.upv.es; Tel.: +34-963877007

Featured Application: The proposed study can be applied to optimize energy consumption, generation and storage in neighbour communities with distributed energy resources.

Abstract: In reliability studies of isolated energy supply systems for residential buildings, supply failures due to insufficient generation are generally analysed. Recent studies conclude that this kind of analysis makes it possible to optimally design the sizes of the elements of the generation system. However, in isolated communities or rural areas, it is common to find groups of dwellings in which micro-renewable sources, such as photovoltaic (PV) systems, can be installed. In this situation, the generation and storage of several houses can be considered as an interconnected system forming a cooperative microgrid (CoMG). This work analyses the benefits that sharing two autonomous installations can bring to each one, from the point of view of reliability. The method consists of the application of a random sequential Monte Carlo (SMC) simulation to the CoMG to evaluate the impact of a simple cooperative strategy on the reliability of the set. The study considers random failures in the generation systems. The results show that the reliability of the system increases when cooperation is allowed. Additionally, at the design stage, this allows more cost-effective solutions than single sizing with a similar level of reliability.

Keywords: microgrid; cooperation; energy; reliability; Monte Carlo; smart energy control

Citation: Roldán-Blay, C.; Roldán-Porta, C.; Quiles, E.; Escrivá-Escrivá, G. Smart Cooperative Energy Supply Strategy to Increase Reliability in Residential Stand-Alone Photovoltaic Systems. *Appl. Sci.* **2021**, *11*, 11723. https://doi.org/10.3390/app112411723

Academic Editor: Mohamed Benbouzid

Received: 10 November 2021
Accepted: 7 December 2021
Published: 10 December 2021

1. Introduction

The increase in energy demand and dependence on fossil fuels [1] has boosted the integration of renewable energies [2]. This has favoured the development of distributed energy resources (DERs) [3], especially in rural or isolated areas [4].

Small size DERs in Spain are usually implemented through the so-called stand-alone photovoltaic (SAPV) energy systems [5–7], which generally include batteries to store energy and supply night demand. These types of systems are currently very attractive in rural areas thanks to the new regulations that have regulated self-consumption since 2019 [8]. In a general case, the energy storage system (ESS) could also be based on renewable energies, as in the case of micro hydroelectric plants or green hydrogen generators [9,10]. The great disadvantage of this type of system is the reliability of supply, since a profitable design usually implies a large number of hours per year of energy not supplied (ENS), increasing indices such as the loss of load probability (LOLP) to inappropriate levels [11]. Due to the high price of ESSs, especially those based on renewable energies, an optimized design of the generation system is of vital importance. In this sense, there are studies that focus on improving the performance of these systems [12–14]. On the other hand, other works are focused on improving the management of generation, demand and storage resources to obtain maximum reliability with the lowest possible cost [15,16].

The reliability of these systems can be improved with support systems such as a diesel generator, but it is necessary to manage the use of each resource with the best possible strategy to minimize diesel consumption, due to its price and environmental impact [17,18]. Thus, many studies have focused on the technical feasibility of these systems and their reliability [19]. A group of the developed studies have focused on the assessment of residential SAPV generation plants connected to the electricity distribution network. These allow the users to sell their energy surpluses or demand extra energy as needed. Another type of study has focused in isolated systems [20,21], which is the only available option in most rural areas. As stated before, the optimal design of these systems is very important [22]. To achieve a good design of these systems there have been a large number of methods developed, as can be seen in some review works [23,24].

One of the greatest measurable benefits of using systems based on renewable energy is the increase in the sustainability of the energy supply [25] and its positive impact against climate change [26]. Furthermore, various studies point to a clear correlation between the integration of these energy systems and economic growth [27–29]. Among the key elements of this type of system, the importance of the ESSs stands out [30]. On the other hand, one of the challenges that is arousing growing interest is the problem of optimizing these systems considering various realistic scenarios. There are multiple objectives that can be set to study this problem, such as energy supply in rural or isolated areas [31,32], losses minimization [33] or reliability improvement [34]. Among these objectives, reliability is one of the least studied and most important aspects in systems based on renewable energies, especially in rural areas, since in many cases the resources are intermittent [30] and difficult to predict [35]. One of the strategies that should be studied to improve the reliability of supply in systems based on renewable energies is the use of configurations following the architecture of a microgrid (MG) with cooperative behaviour [36], also called CoMG. This configuration can benefit the entire system in the different objectives discussed [37].

In summary, there are numerous papers that address the optimization of energy management based on renewable sources. However, they generally do not analyse the problem from the point of view of reliability and the guarantee of achieving a certain value in continuity of supply [6,9,17,18,20,23,24,32]. One of the few papers that includes reliability as a goal to design the size of an autonomous system with renewable energy is [34]. It optimizes the size of a system with three sources (solar, wind and tide) and an ESS. Reliability is studied from an annual curve for each resource (solar radiation, wind speed, water speed and demand) that reflects the mean values of 16 years. The results depend on the costs assigned to each system, but the advantage of a cooperative operation is not studied, which is the main objective of this paper. This is the main gap identified in the literature, since the published studies do not include, at the same time, a cooperative management strategy including random failures in the system, reliability optimization and optimal design of the facilities.

In a previous study [38], the authors developed a method to evaluate the reliability of an SAPV energy system in an isolated case using an SMC simulation methodology, which provides very useful information to improve the design of these systems. The methodology developed in this study is applicable with various objectives. On the one hand, the method allows us to propose an optimal design of the system to achieve the desired reliability. On the other hand, it allows us to study the reliability with different designs, which facilitates the analysis of the necessary support system (generally, a diesel generator). Finally, the parameterized result obtained with the developed method allows for a complete analysis of the system to be designed, controlling not only aspects related to the reliability of the supply but also the use of the energy generated and the sensitivity analysis of the system to various disturbances. It is important to note that this study aims at optimizing the size of the components of a single facility.

As indicated in the conclusions of this study, the authors consider that it is necessary to study the reliability of this type of system when several installations are connected in an MG. Going one step further, it is very interesting to analyse the impact on the reliability and optimal design of these systems of CoMGs. This will further improve facility designs and assess the reliability of the system as a whole, optimizing energy use and minimizing the use and size of support systems. Thus, in this work, two neighbouring houses whose demand throughout 2019 is known have been selected and a cooperative management strategy has been developed. The main objective of the presented study is to evaluate the reliability of the system and obtain an optimal design that improves the results of the individual design for each participant. The 2020 data have not been considered adequate after some analysis due to the pandemic situation, especially considering that an SMC method will be used to simulate multiple scenarios and draw reliable global conclusions. This study is the basis for evaluating the design and reliability of MGs in which each installation has its own SAPV energy system and its ESS and a connection to the MG. One type of these CoMGs is called a virtual power plant (VPP) in the literature. There are numerous works on VPPs. For an exhaustive review, works such as [39] can be consulted. However, these works focus in optimizing the management strategy of the different resources, which should only be conducted after an optimal design.

In this work, starting from a design based on the method described in [38], a simple cooperative management strategy has been developed and the reliability of the system has been evaluated using an SMC simulation approach. The method used allows the inclusion of random faults in the generation systems, which is one of the most overlooked aspects in reliability studies. The parameterized analysis of the results shows the potential to improve the design considering the cooperative management of the MG, which allows the minimization of costs, the minimization of the size and usage of support systems and a significant improvement in the reliability of the system as a whole. Thus, the main novelty of this work is the possibility of achieving an optimized design of a set of interconnected facilities, studying their reliability through a parameterized analysis. With the results of the analysis, it is possible to select the most appropriate design alternative according to the selected objectives.

The specific objectives that have been addressed in this work and that represent the novelty of the study are detailed below.

1. Study the reliability of a set of two dwellings with their respective SAPV energy systems when operating with a cooperative strategy.
2. Improve the design of all the systems considering global reliability.
3. Analyse the impact of the cooperative strategy on the reliability and design of the systems.

In addition, as another novel aspect in the procedure used in this study, random failures in the generation system are considered to analyse supply reliability, in addition to the use of demand and generation curves with a random combination so that the result is consistent and robust. These random failures are generated following typical distributions extracted from scientific works.

With the presented method the authors try to address the gap identified in the literature of the optimal design and management in microgrids considering costs and reliability as the main targets. The novelty of the presented work relies on the practical application of a simple strategy to a rural area considering real data and random failures of the system to optimise the design of the components and the management with a cooperative strategy. This problem including cooperation between both systems represents a challenge that takes place in many isolated areas in the developed countries. The results of the method provide an economic benefit from the planning stage of the microgrid and an increase in the reliability of the participants' energy supply, which is one of the main gaps in scientific publications.

The rest of the article has been structured as follows. Section 2 shows the materials and methods, summarizing the individual design of the dwellings using the method developed in [38] and proposing the cooperative management strategy to be used for the SMC in the selected scenarios. Section 3 shows the results of the SMC simulation comparing the system without interconnection with the system as an MG. Next, Section 4 shows the discussion of the results. Finally, Section 5 shows the conclusions of the study.

2. Materials and Methods

For this work, two neighbouring dwellings (D1 and D2) have been selected whose hourly demand has been registered for a whole year. In each of them, an SAPV energy system with an ESS made of batteries has been proposed, using an energy analysis, as described in [38]. Subsequently, using the SMC simulation method described in [38] the initial design has been optimized to achieve an LOLP around 2.6% (approximately between 200 and 240 h per year with energy supply from a diesel generator) in each dwelling.

The SMC consists of a simulation of 100 iterations of the behaviour of the system, each of them consisting of one year (8760 h). For each iteration of the study, an hourly sequence of the demand in each house during one year is used, randomly obtained from the real values measured during 2019.

The solar radiation (and therefore the power production per unit in the PV panels) is assumed to be the same for both dwellings, since they are close to each other. At each iteration, the daily radiation curve is randomized from the real values recorded during 2019.

Additionally, in each iteration, the failures of the generation system are also randomized, using a failure rate per year (λ_c). As explained in [40], the time to failure (TTF) of the PV plant and the mean time to repair (TTR) are modelled using an exponential and a Rayleigh distribution, respectively. Then, TTF and TTR are randomly generated with the inverse transform method [41].

From this simulation the mean values of the indices to be estimated are obtained: energy not used (ENU), hours with energy not used (HNU), ENS and LOLP. These indices are obtained every time for each dwelling individually and for the complete system.

The results obtained with the SMC simulation of each dwelling constitute the S1 scenario described in Section 2.1, which corresponds to the reliability of the system working as isolated dwellings. In this initial scenario, there is no interconnection between both houses. This scenario has been compared with a set of scenarios called $S2_x$ (for different values of x), which consist of cooperative scenarios in which both dwellings are considered interconnected in an MG and can share their resources, as explained in Section 2.2.

The main objective of defining this set of scenarios is to optimize the cooperation strategy so that both houses can benefit in a similar way. Figure 1 shows a diagram of the scenarios considered.

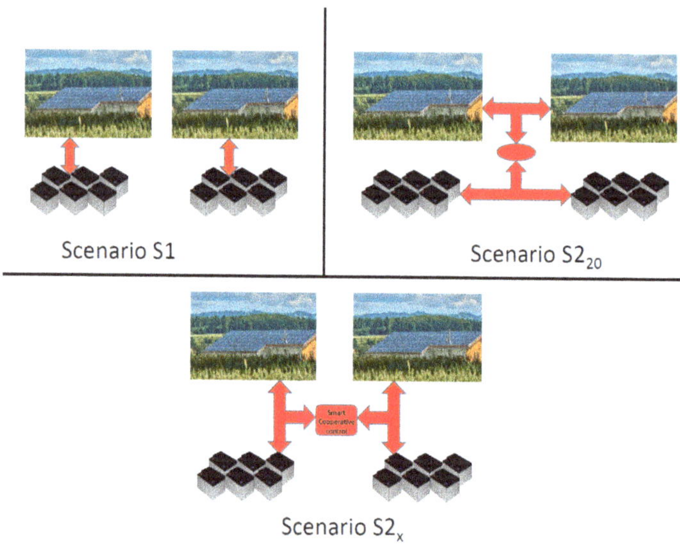

Figure 1. Description of the scenarios considered.

2.1. Scenario S1

In scenario S1, the energy flows of each house are simple. In batteries, the state of charge (*SOC*) must be continuously monitored. All batteries have an SOC_L up to which discharge is allowed. An SOC_M can also be defined as the maximum *SOC* that should not be exceeded, which for this study is set at $SOC_M = 100\%$. If generation is greater than demand, the surplus energy is available to charge the ESS with an efficiency η_c corresponding to the charging efficiency. If the ESS cannot be charged because it has reached SOC_M, the available energy that could have been generated becomes a new amount of ENU. On the other hand, if generation is lower than demand, the deficit of energy being actually demanded can be supplied from the ESS with the efficiency η_d corresponding to the discharging efficiency. Under these conditions, if the ESS cannot be discharged because it reaches SOC_L, then there is a certain amount of ENS that will affect the studied indices. The operation of this scenario and the different situations that can take place are described in sufficient detail in [38]. Figure 2 shows the different energy lows and efficiencies in this scenario.

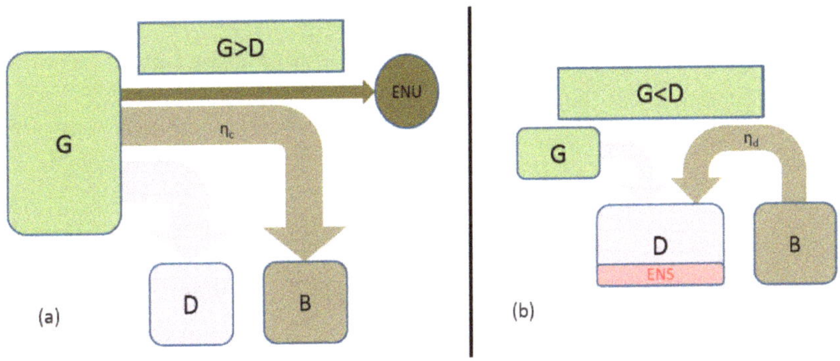

Figure 2. Possible energy flows in S1 when generation is greater than demand (**a**) and when generation is lower than demand (**b**).

2.2. Scenarios $S2_x$

$S2_x$ scenarios represent an actual CoMG situation. Each dwelling tries to be self-sufficient with its generation and its ESS, but, under certain conditions, each house can provide energy to the other when needed if it has enough energy surplus. To prevent one dwelling from incurring significant reliability losses when helping the other, a simple strategy has been established: dwellings can share energy if the stored energy is above a threshold, but they stop cooperating if it falls below that limit. Thus, this threshold is the key parameter of the proposed strategy. If the ESS is made up of batteries, the strategy entails allowing power sharing if $SOC > SOC_C$, with SOC_C being the *SOC* value set as the threshold for power sharing. Each scenario will be indicated as an $S2_x$, where $x = SOC_C$ (%), using x as the key parameter of the study. If $x = SOC_C = SOC_L$ (which is the lower limit) is chosen, the scenario is equivalent to two houses with shared resources, without a cooperative strategy. This case will be called a shared scenario. In this work, $SOC_L = 20\%$ has been set, so the $S2_{20}$ scenario corresponds to the case of the shared scenario but without an intelligent cooperative control, as depicted in Figure 1. In this case, there would be no cooperative strategy and the system would behave as if the resources of both houses were joined together and shared freely by both. For all $S2_x$ scenarios, the aforementioned indices have been parameterized as a function of $x = SOC_C$ between $20\% \leq SOC_C \leq 40\%$.

It is easy to verify that the situation in $S2_{20}$ produces unequal results: the home with the worst LOLP and the highest ENS has great benefits, but the one with the best starting indices worsens. This situation would not be admissible for the dwelling with a better initial situation in terms of LOLP and ENS. Therefore, it is necessary to differentiate between this particular case and two dwellings with a cooperative strategy to try to benefit both ($S2_x$ scenarios). Therefore, the $S2_x$ scenarios are parameterized as a function of x, studying various cases depending on the cooperation capacity of the two dwellings, as described below.

In $S2_x$, the energy flows are slightly more complex, since under certain conditions an energy flow may appear from one dwelling (either from the generation system or ESS) to the other. In these crossed energy flows, the transmission efficiency η_t has to be applied. For the explanation of the method, a superscript j will be used to indicate the dwelling to which each magnitude refers. For the calculation of the energy flows in these scenarios, if generation in a dwelling (Dj) is greater than demand, the surplus is available to charge its own ESS with the charging efficiency η_c. If $SOC^j > SOC_C^j$, the excess between generation and demand would be available to supply energy to the other dwelling with the transmission efficiency η_t. Additionally, the energy stored in the battery of Dj between SOC^j and SOC_C^j could be offered to the other house, with an efficiency $\eta = \eta_d \cdot \eta_t$, where η_d is the discharging efficiency.

The energy stored in the battery of Dj at each moment is $B^j = SOC^j \cdot B_r^j$, where B_r^j is the nominal capacity of housing battery j.

Therefore, the following values are obtained:

- $B_M^j = SOC_M^j \cdot B_r^j$, maximum energy that can be available in battery of Dj.
- $B_C^j = SOC_C^j \cdot B_r^j$, value of energy stored in battery of Dj, below which it is not allowed to export energy to help the other dwelling.
- $B_L^j = SOC_L^j \cdot B_r^j$, minimum value of stored energy that must be maintained at all times to avoid damage to the battery.

The energy that will finally be put into circulation will be the minimum value between the energy available to help the other home and the maximum value required by the other home.

If the generation capacity exceeds the sum of demanded energy, storage capacity and energy transmitted to the other home, there will be an excess of generation capacity that will not be used, resulting in a certain amount of ENU.

Figure 3 shows the energy flows in $S2_x$.

If the generated energy is lower than demand, some demanded energy should be supplied with the energy stored in the ESS, with the efficiency η_d. If the energy available in the ESS ($B^1 - B_L^1$) is not enough to cover the demand, a situation with a need for energy from the transmission system is reached. To improve the reliability of the supply, it is also considered the energy needed in the ESS if it does not reach the SOC_C, which is $B_C^1 - B^1 \geq 0$.

When $SOC > SOC_C$ energy is not demanded from the transmission system, since this energy is subjected to various actions with yields lower than 1 and, therefore, these transfers must be limited.

If the sum of the own generation, the energy available in the battery and the energy received from the cooperative system is not enough to cover the demand, there will be an ENS, which will be counted and will affect the studied indices.

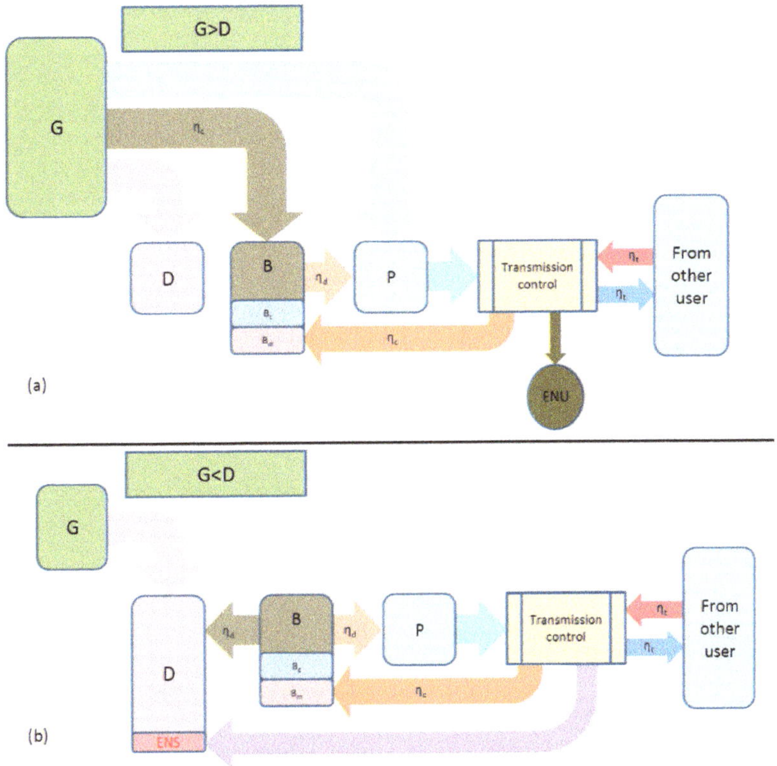

Figure 3. Possible energy flows in $S2_x$ when generation is greater than demand (**a**) and when generation is smaller than demand (**b**).

From the demand and generation curves of each house, with the random failures generated and using a step of 1 h, the following set of equations describes the cooperative strategy to apply. The initial data needed for each dwelling are B^j_M, B^j_C, B^j_L, η_c, η_d, η_t. Then, for each step i, the values of G^j_i, D^j_i, B^j_i must be obtained. Next, the surplus generation U^j_i is calculated for each dwelling using Equation (1)

$$U^j_i = G^j_i - D^j_i \tag{1}$$

where U^j_i is the usable energy that represents the surplus energy generation. If $U^j_i > 0$, there is a net surplus generation, so the next equations must be used

$$B'^j_i = B^j_i + U^j_i \cdot \eta_c \tag{2}$$

$$E^j_i = B'^j_i - B^j_M \geq 0 \tag{3}$$

$$B''^j_i = \min\left\{B^j_i,\ B^j_M\right\} \tag{4}$$

Otherwise, when $U^j_i < 0$, Equations (2)–(4) must be substituted by the following expressions, since there is a net generation deficit

$$B'^j_i = B^j_i + \frac{U^j_i}{\eta_d} \tag{5}$$

$$E_i^j = B''^j_i - B_L^j \leq 0 \tag{6}$$

$$B''^j_i = \max\left\{B'^j_i,\ B_L^j\right\} \tag{7}$$

After these intermediate calculations have been computed, Equation (8) is used to calculate N_i^j, which represents the energy needed to supply the demand and to increase the energy stored up to B_C^j.

$$N_i^j = B_C^j - B'^j_i \geq 0 \tag{8}$$

The available energy to be sent to another participant, P_i^j, also called pool energy, can be derived from Equation (9) as

$$P_i^j = E_i^j + \left(B''^j_i - B_C^j\right) \cdot \eta_\mathrm{d} \geq 0 \tag{9}$$

With these values, the following equations must be used.

$$T'^{j\to k}_i = \min\left\{\frac{N_i^k}{\eta_\mathrm{t}},\ P_i^j\right\} \tag{10}$$

$$T_i^{k\to j} = T'^{k\to j}_i \cdot \eta_\mathrm{t} \tag{11}$$

where $T'^{j\to k}_i$ is the energy that will be sent from Dj to Dk, and $T_i^{k\to j}$ is the energy received by Dj from Dk.

At the end of step i, the energy stored in the battery, B_i^{+j}, in Dj is calculated using Equation (12)

$$B_i^{+j} = B''^j_i - \frac{T'^{j\to k}_i}{\eta_\mathrm{d}} + T_i^{k\to j} \cdot \eta_\mathrm{c} + E_i^j \cdot \eta_\mathrm{c},\ \mathit{limited\ to}: B_L^j \leq B_i^{+j} \leq B_M^j \tag{12}$$

For the next step $i+1$, the battery level will be updated as $B_{i+1}^j := B_i^{+j}$. Finally, the following two parameters are computed for each dwelling to evaluate the results of the simulation

$$ENU_i^j = E_i^j - T'^{j\to k}_i \geq 0 \tag{13}$$

$$ENS_i^j = T_i^{k\to j} - E_i^j \geq 0 \tag{14}$$

2.3. Summary of the Proposed Method

Using the definitions and equations presented in this section, the steps used in the presented work are explained below.

The first stage was to collect the data of the facilities for a whole year. Then, with these data, the method described in [38] was used to obtain the optimal design for both facilities. As a result of this initial design, scenario S1 was simulated. As aforementioned, this scenario consists of two isolated facilities, so there is no possibility to interact or cooperate.

The simulation of a scenario is developed by performing 100 iterations of the yearly operation of both facilities to obtain the average reliability indices once the dispersion has been stabilized. These results are stored to compare all scenarios at the end of the study. This process is based on an SMC simulation to obtain the average reliability indices.

After analysing this scenario, the set of scenarios $S2_x$ for $x \in [20\%, 40\%]$ are studied. The first scenario in this set is $S2_{20}$, in which the resources are shared between both facilities. After completing the 100 iterations (the SMC simulation) and calculating the reliability indices, the value of x is increased in 5% and a new scenario is proposed.

Once all scenarios have been studied, the analysis of the reliability indices provides useful information about the systems. This information is discussed and some conclusions are drawn about the optimal design and management of the resources for this set of facilities. The proposed method is summarized in Figure 4.

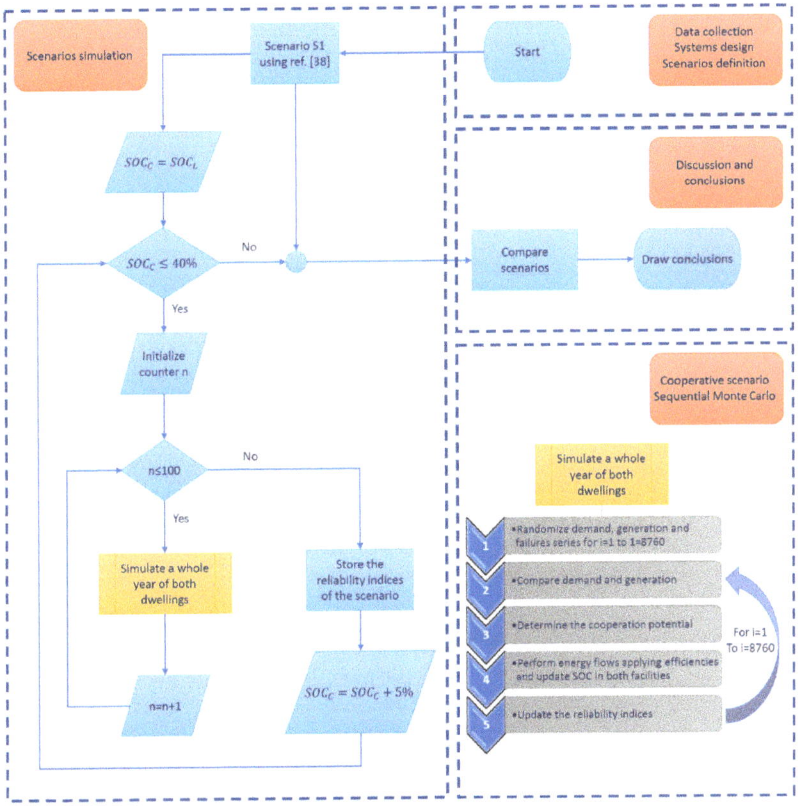

Figure 4. Stages of the proposed method.

3. Results

The proposed method was applied to two real dwellings located in Valencia (Spain). The hourly demand curves were recorded during 2019 (avoiding the pandemic situation of 2020, as mentioned above). Likewise, a solar photovoltaic generation curve was registered in the GEDERLAB, a real facility next to the dwellings in which the generation and management of various renewable energy sources is studied. The recorded data was randomized swapping nearby days to develop the SMC simulation, so that the demand and generation curves maintained their seasonal patterns, but irregular situations such as a special event in a dwelling or a cloudy day did not match in each simulation. Figure 5 shows the typical patterns of demand for both dwellings and generation in each season of the year in per unit values.

Using the rules explained from Equation (1) to Equation (14), the SMC simulation was carried out. In each simulation, the results of all scenarios were evaluated. The simulation was repeated 100 times, checking that the average indices stabilized in order to analyse the results and draw conclusions.

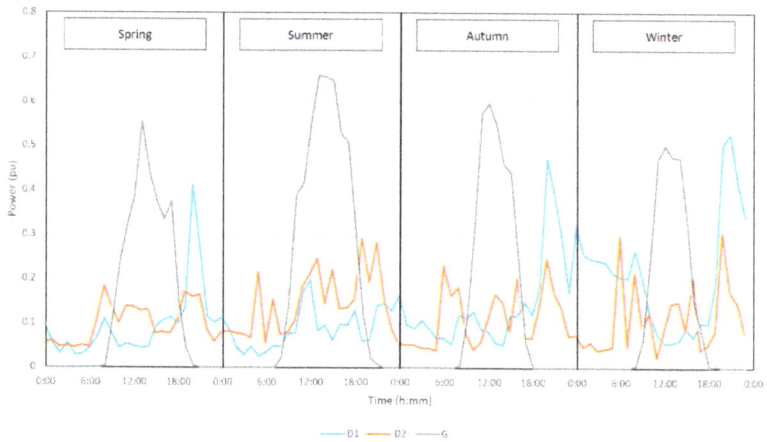

Figure 5. Demand profile of each dwelling and generation profile for every season in per unit values.

To simulate the different scenarios, we started from the optimal design of both dwellings shown in Table 1, obtained after analysing the demand and generation of each dwelling using the method described in [38].

Table 1. Generation, demand and design data for both dwellings.

Variable	Value (D1)	Value (D2)	Description
$P_d^j(peak)$	3 kW	3.3 kW	Maximum demanded power
E_y	3.36 MWh	3.4 MWh	Annual energy consumption
η_c, η_d	0.9	0.9	Efficiency of Li-ion battery (charge and discharge)
SOC_L	20%	20%	Minimum allowed SOC
SOC_C	20% ÷ 40%	20% ÷ 40%	Minimum SOC considered to allow cooperation in scenario 2_x
Failure rate λ_c	2 f/yr	2 f/yr	Exponential distribution
Repair time	24 h	24 h	Rayleigh distribution
PV_{peak}^j	6.5 kW	5.5 kW	Calculated rated power in PV panels in dwelling j, optimized with MCS method in S1
B_r^j	23 kWh	18 kWh	Rated battery capacity in dwelling j obtained in S1

3.1. Scenario S1

Scenario S1 consists of optimizing the PV_{peak} and B_r values of each dwelling, operating independently, to achieve an approximate LOLP of 2.6% in each case. The resulting values are shown in Table 1.

The simulation of 100 years in this situation allows obtaining the stationary values of the parameters shown in Figure 6, which represent the indices expectation. As shown in Figure 6, the design has been adjusted so that both dwellings have a similar LOLP value, although the ENS is not the same due to the differences in the registered demand curves.

Although the radiation curves are identical for both dwellings, since they are neighbouring houses, the registered demand curves are different. For this reason, it is expected that the cooperative strategy can benefit both participants, thus improving the reliability indices. Figure 7 shows the generation and demand profiles obtained for a complete week in both dwellings with the proposed design.

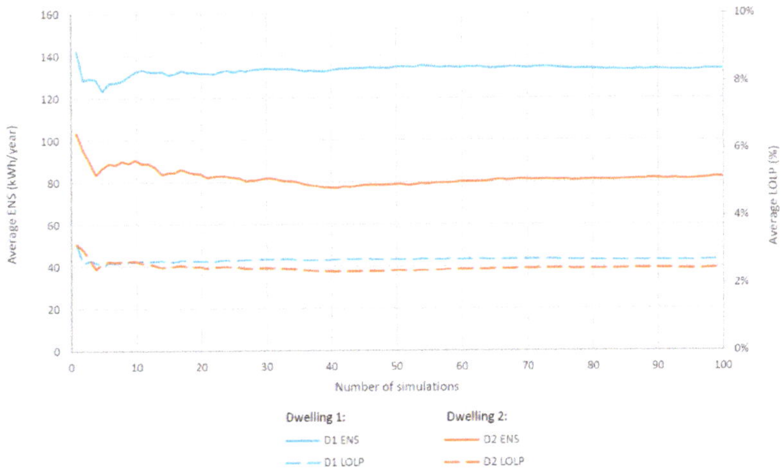

Figure 6. SMC results in Scenario S1 for the most representative indicators (ENS and LOLP).

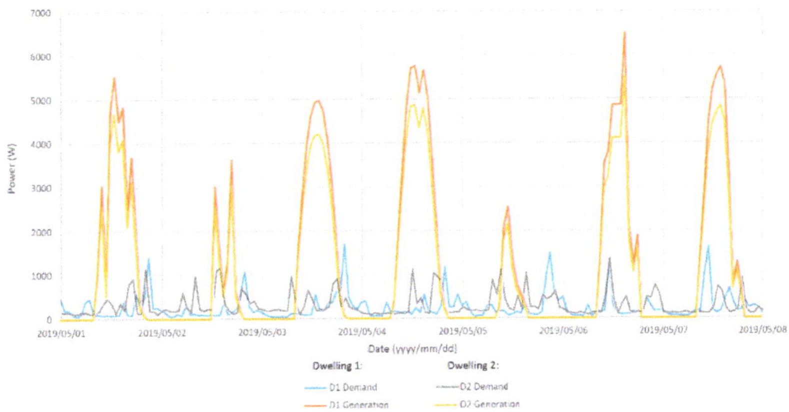

Figure 7. Generation and demand profiles during a week with the proposed design.

3.2. Scenario $S2_x$

A cooperative strategy is studied in each $S2_x$ scenario. Each dwelling supports the other as long as its battery is above SOC_C. Between this value and $SOC_L = 20\%$ (minimum value considered as a limit), the dwelling reserves its energy for its own use. This strategy avoids some interruptions in one of the dwellings, while it allows them to help the other in some periods of time.

With this strategy, it is also possible to take advantage of the energy surplus of one installation to charge the other's ESS, which would reduce the overall ENU.

The key parameter for the study of this cooperative strategy is the value of SOC_C. If $SOC_C = SOC_L$, then all of the energy available in each dwelling can be shared with the other. This situation can be beneficial in global terms, but if the reliability of each dwelling or the consumption pattern is different, this shared strategy produces unequal benefits. The dwelling with the worst performance is greatly benefited, while the other may have a much lower improvement. This situation seems clearly unfair and it would be rejected by the dwelling with the best starting situation, since it would encourage dwellings to try to design their facilities with fewer resources and then take advantage of others' resources. Therefore, starting from this shared scenario called $S2_{20}$, the value of SOC_C will be gradually increased to analyse the improvements obtained in each dwelling and find the best situation for both.

Table 2 shows the global results of both dwellings and the individual results of $S2_{20}$. It is observed that this scenario produces better overall results than S1.

Table 2. Scenario S1 vs. scenario $S2_{20}$.

	ENU (Wh)	HNU (h)	ENS (Wh)	LOLP (%)
D1 alone (S1)	4,552,022.19	1825	133,083	2.70%
D2 alone (S1)	3,290,158.96	1584	80,284	2.38%
D1 shared ($S2_{20}$)	4,283,190.49	1611	100,543	2.18%
D2 shared ($S2_{20}$)	3,252,527.15	1514	67,712	2.13%
S1 (global)	7,842,181.15	1705	213,368	2.54%
$S2_{20}$ (global)	7,535,717.64	1563	168,255	2.15%
$S2_{20}$ improvement compared to S1 (%)				
D1	5.91	11.75	24.45	19.09
D2	1.14	4.42	15.66	10.58
Average	3.91	8.34	21.14	15.10

As the table shows, there are significant improvements in reliability indices. However, dwelling D2, which was the one with the best initial results, is less benefited than D1. D2 is being more helpful to D1 than the other way around, which means that the situation is not fair, since the ideal scenario would be that both contributed in a similar way to improve the system.

That is why the $S2_{30}$ scenario is simulated, in which every dwelling reserves an energy reserve to supply its own demand, so that it stops relying on the other dwelling when $SOC_C = 30\%$. The results of this scenario compared to scenario S1 are shown in Table 3.

Table 3. Scenario S1 vs. scenario $S2_{30}$.

	ENU (Wh)	HNU (h)	ENS (Wh)	LOLP (%)
D1 alone (S1)	4,560,010.28	1828	131,561	2.65%
D2 alone (S1)	3,292,706.47	1585	82,274	2.46%
D1 cooperative ($S2_{30}$)	4,292,284.49	1617	106,547	2.24%
D2 cooperative ($S2_{30}$)	3,260,417.70	1519	66,109	2.09%
S1 (global)	7,852,716.75	1706	213,835	2.55%
$S2_{30}$ (global)	7,552,702.19	1568	172,656	2.16%
$S2_{30}$ improvement compared to S1 (%)				
D1	5.87	11.52	19.01	15.59
D2	0.98	4.18	19.65	15.22
Average	3.82	8.11	19.26	15.41

In this scenario, the results obtained by both dwellings are very similar. It would not be a good idea to arbitrarily raise the SOC threshold that has been set as a limit to stop sharing resources, that is SOC_C, since then the support between the dwellings would decrease. For example, setting $SOC_C = 35\%$, scenario $S2_{35}$ is simulated, the results of which are shown in Table 4, compared again with those of scenario S1.

As it can be seen, although the total results are very similar to the previous case, dwelling D2 is now the most favoured. Finally, if SOC_C is further increased to a value of $SOC_C = 40\%$, the results worsen, as seen in Table 5, where scenario $S2_{40}$ is compared with scenario S1.

Table 4. Scenario S1 vs. scenario $S2_{35}$.

	ENU (Wh)	HNU (h)	ENS (Wh)	LOLP (%)
D1 alone (S1)	4,556,147.37	1827	131,088	2.64%
D2 alone (S1)	3,289,774.14	1586	79,574	2.33%
D1 cooperative ($S2_{35}$)	4,288,604.12	1615	107,872	2.26%
D2 cooperative ($S2_{35}$)	3,255,462.81	1515	61,143	1.89%
S1 (global)	7,845,921.51	1706	210,662	2.49%
$S2_{35}$ (global)	7,544,066.93	1565	169,015	2.08%
$S2_{35}$ improvement compared to S1 (%)				
D1	5.87	11.60	17.71	14.46
D2	1.04	4.47	23.16	18.82
Average	3.85	8.29	19.77	16.50

Table 5. Scenario S1 vs. scenario $S2_{40}$.

	ENU (Wh)	**HNU (h)**	**ENS (Wh)**	**LOLP (%)**
D1 alone (S1)	4,558,096.21	1828	136,035	2.76%
D2 alone (S1)	3,292,984.25	1587	83,180	2.47%
D1 cooperative ($S2_{40}$)	4,296,492.29	1618	116,710	2.42%
D2 cooperative ($S2_{40}$)	3,261,013.86	1519	66,528	2.08%
S1 (global)	7,851,080.46	1707	219,215	2.61%
$S2_{40}$ (global)	7,557,506.15	1568	183,239	2.25%
$S2_{40}$ improvement compared to S1 (%)				
D1	5.74	11.47	14.21	12.47
D2	0.97	4.29	20.02	15.64
Average	3.74	8.13	16.41	13.97

Therefore, the $S2_{30}$ scenario offers the most balanced results in terms of the improvements obtained by both dwellings. After applying the proposed method, the study was completed with a sensitivity analysis shown in Section 4.

4. Discussion

To see the evolution of the reliability of the dwellings for different values of SOC_C, a sensitivity analysis was carried out with SOC_C values between $SOC_L = 20\% \leq SOC_C < 60\%$, simulating each of the resulting scenarios. By increasing the value of SOC_C, the dwellings become more independent and the advantage of cooperation is lost, which is why it has been simulated in the aforementioned interval.

Figure 8 shows the improvement of LOLP in percentage compared to S1 of both dwellings in the simulated scenarios of this sensitivity analysis.

As depicted in Figure 8, with small values of SOC_C, D1 improves much more than D2. This difference decreases when SOC_C increases, and for values close to SOC_C = 30%, both houses achieve a similar benefit.

The curves represented in Figure 9 show the LOLP of each dwelling in this case (scenario $S2_{30}$), which is shown to be the best of the studied cases, compared to the results of scenario S1.

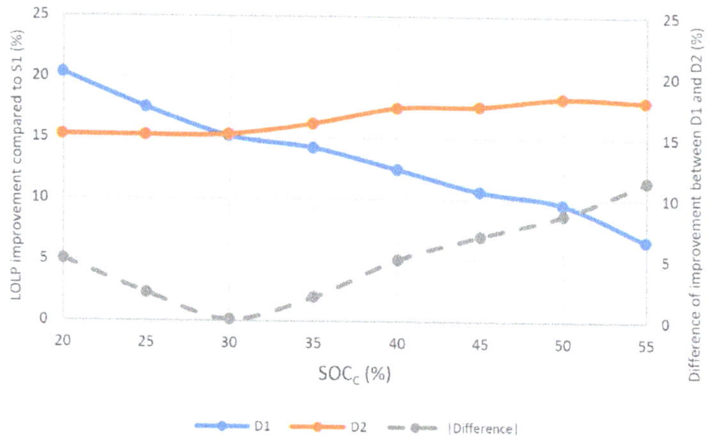

Figure 8. Improvement in terms of LOLP for each dwelling in scenarios $S2_x$ compared to S1.

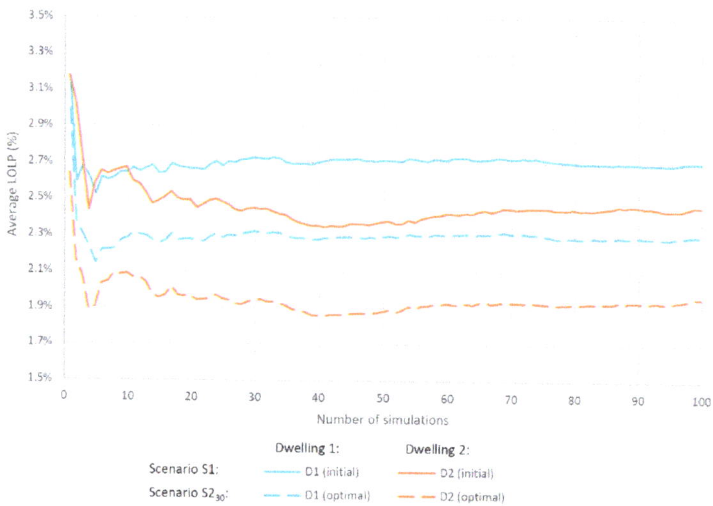

Figure 9. LOLP of each dwelling in scenarios S1 and $S2_{30}$.

In addition, it should be noted that not only does the cooperative strategy represent an improvement compared to the S1, but it also improves the results of the shared scenario $S2_{20}$, as shown in Figure 10. This analysis solves one of the research gaps identified in the literature, as it proposes a simple cooperative strategy based on the key parameter SOC_C. With this strategy, the system obtains better results than in other scenarios (shared and isolated), which correspond to other microgrid configurations in the literature (since most studies, such as [34], are exclusively focused on stand-alone). In addition, these results have a substantial impact to be taken into account, since they have been obtained using real data and proposing random failures in the system.

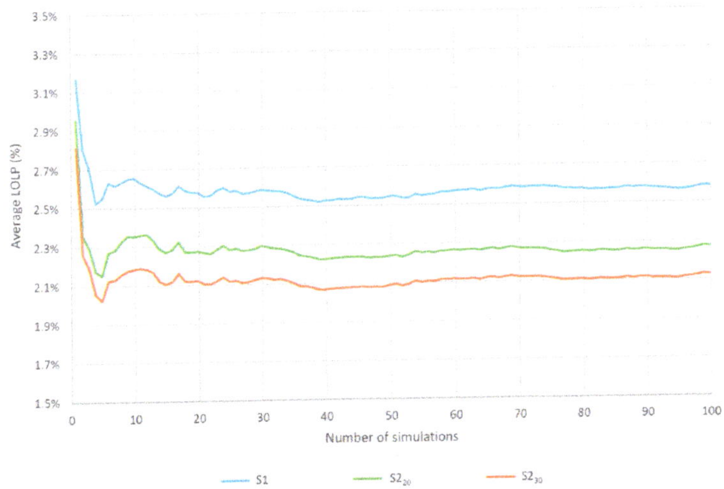

Figure 10. LOLP of each dwelling in scenarios S1, $S2_{20}$ and $S2_{30}$.

The results obtained in this case study show that for a set of demand and generation curves, the most equitable strategy can be proposed by setting the parameter of SOC_C at a certain value that can be determined with the proposed method. For this case study, this value was 30%, which made it possible to achieve better reliability values in both dwellings than those obtained without using cooperative strategies, while providing balanced improvements. Compared to most of the literature works (such as [22]), this is one of the key contributions of this study, since the cooperative strategy allows performance optimization and reliability improvements.

In addition to the presented case study, the described method would allow us to obtain an optimal design in different situations. One of the situations that could be addressed with the proposed method would be that of consumers who had generation curves with different patterns. This can take place when the geographical location of the considered consumers receives different radiation levels or when there is a different generation mix in the facilities (for example, including a wind turbine in one of them). The fact of having different generation patterns, in addition to having different load curves, further justifies the use of this cooperative strategy, since there is a greater probability to have one of the dwellings with an energy surplus when the other is in need of additional generation.

A case of special interest to study of this type of strategies is that of Virtual Power Plants (VPPs), since in the set of generation and demand curves aggregated in a VPP there can be a great variety of scenarios that can provide great benefits derived from the use of cooperative strategies.

In the developed case study, one of the implicit benefits obtained when planning the facilities to use cooperative strategies is the net reduction in installation costs due to the optimized use of resources. This has been carefully addressed in this work compared to the literature by adding random faults in the system to simulate the repair times and include their impact on the reliability indices (as opposed to most studies, such as [11]). Since both facilities share resources and they use them cooperatively, it is possible to obtain results similar to those obtained in S1 with a lower amount of installed resources. S1 corresponds to a scenario in which there is no interconnection between the two dwellings, so the cooperative strategy is not needed. For this reason, this scenario has been completed using the method described in [38]. The new proposal allows for the generalization of that method for a set of multiple prosumers in a single microgrid. From the improvement in this method it can be derived that to obtain the same results as in scenario $S2_{30}$ without sharing resources, it would be necessary to install a greater ESS in each facility. In scenario

$S2_{30}$, the reliability parameters obtained are an ENS of 106,547 kWh and 66,109 kWh and a LOLP of 2.24% and 2.09% for D1 and D2, respectively. In order to reach these values for both dwellings in S1, the value of B_r^j would have to be as shown in Table 6. This allows quantifying of the benefits of the proposed method compared to previous studies for the selected case study.

Table 6. Initial design of batteries for scenario S1 to obtain the same results as in scenario $S2_{30}$.

	B_r^j (kWh)	LOLP (%)	Extra Capacity Needed (%)
D1	26,000	2.21	13.04
D2	20,000	2.02	11.11

Finally, together with the economic benefit obtained thanks to the described method, it is worth highlighting the environmental benefit of using fewer resources and needing less support from conventional systems, since in this type of dwellings diesel generators are commonly used to supply energy in the hours in which the SAPV system and the ESS are not able to cover the demand.

5. Conclusions

In this study, a method to analyse and improve the reliability of energy supply in two dwellings with their SAPV system and their ESS when they use a cooperative strategy has been proposed and validated. The method is based on the use of surplus generation and storage of one facility to supply energy to the other facility when needed, fixing a minimum amount of energy to stop cooperating in order to avoid losing energy needed to supply its own demand, defined by means of the key parameter SOC_C. To test the method, it has been applied to two real neighbouring dwellings, using real generation and demand curves. The algorithm is based on the use of a random SMC simulation to find the stable values of the selected reliability indices. As the simulations show, the results of the cooperative strategy are better than the individual results in terms of reliability, but the improvement in both dwellings can be unequal. The method has proven to be useful to find the most even strategy adjusting the level of the minimum SOC needed to start cooperating.

It is important to note that the data were restricted to 2019 due to the anomalous demand pattern shown during the pandemic situation, but the study can be repeated with any other data. In fact, the method shown in this work could be applied to a larger scale case study, i.e., to two neighbouring communities, with larger demand needs and a greater generation mix. As a future work, the authors propose to explore the application of a strategy such as the one proposed here for virtual power plants (VPP), since this would allow the use of more demand curves and different generation curves for each consumer, leading to higher differences between the analysed scenarios and probably to higher chances to improve the global reliability. Additionally, studying the economic impact of the proposed method is a research line that the authors propose to include in future studies.

Author Contributions: Conceptualization, C.R.-B. and C.R.-P.; Data curation, C.R.-B., E.Q. and G.E.-E.; Formal analysis, C.R.-P.; Funding acquisition, C.R.-P. and G.E.-E.; Investigation, C.R.-B. and C.R.-P.; Methodology, C.R.-B. and C.R.-P.; Project administration, C.R.-P. and G.E.-E.; Resources, G.E.-E.; Software, C.R.-B.; Supervision, C.R.-P.; Validation, C.R.-B., E.Q. and G.E.-E.; Visualization, C.R.-P. and E.Q.; Writing—original draft, C.R.-B. and C.R.-P.; Writing—review and editing, C.R.-B. and G.E.-E. All authors have read and agreed to the published version of the manuscript.

Funding: This work was supported by Conselleria de Educación, Investigación, Cultura y Deporte (Grant No. AICO/2019/001).

Institutional Review Board Statement: Not applicable.

Informed Consent Statement: Not applicable.

Conflicts of Interest: The authors declare no conflict of interest.

Abbreviations

B	Energy stored in battery at the beginning of the step
B^+	Energy stored at the end of the step
B', B''	Intermediate results related to stored energy
B_C	Minimum level of energy in the battery to allow cooperating
B_L	Minimum accepted level of battery
B_M	Maximum battery capacity in kWh
B_r	Rated battery capacity
D	Demanded energy
Dj	Dwelling j, j ∈ {1,2}
E	Energy surplus. If $E \geq 0$, there is an energy surplus. If $E < 0$, there is an energy deficit.
ENS	Energy not supplied
ENU	Energy not used
ESS	Energy storage system
E_y	Annual energy consumption
G	Generated energy
HNU	Hours with energy not used
i	Subscript, referring to each step
j	Superscript, referring to each participant
LOLP	Loss of load probability index
N	Energy needed to supply the demand and to increase the energy stored to B_C
P	Available energy to be sent to other participant
$P_d(peak)$	Maximum demanded power in dwelling j
PV_{peak}	Rated power installed in the photovoltaic panels
SAPV	Stand-alone photovoltaic system
SMC	Sequential Monte Carlo
SOC	State of charge
SOC_C	Minimum level of state of charge to allow cooperation
SOC_L	Minimum allowed level of state of charge
SOC_M	Maximum allowed level of state of charge
S1	Scenario 1 (base scenario without sharing resources)
$S2_x$	Scenario 2_x (cooperative scenario)
TTF	Time to failure
TTR	Time to repair
$T'^{j\to k}$	Intermediate result related to energy to be transferred from user j to user k
$T^{k\to j}$	Energy received in user j from user k
U	Usable surplus. If $U \geq 0$, the demand is covered. If $U < 0$, the demand is not covered
VPP	Virtual power plant
η_c	Efficiency of battery charging
η_d	Efficiency of battery discharging
η_t	Efficiency of energy transmission
λ_c	Photovoltaic panel array and battery controller failure rate per year

References

1. Al-Maamary, H.M.S.; Kazem, H.A.; Chaichan, M.T. The impact of oil price fluctuations on common renewable energies in GCC countries. *Renew. Sustain. Energy Rev.* **2017**, *75*, 989–1007. [CrossRef]
2. Lasseter, R.H. Smart Distribution: Coupled Microgrids. *Proc. IEEE* **2011**, *99*, 1074–1082. [CrossRef]
3. Bouffard, F.; Kirschen, D.S. Centralised and distributed electricity systems. *Energy Policy* **2008**, *36*, 4504–4508. [CrossRef]
4. Woldeyohannes, A.D.; Woldemichael, D.E.; Baheta, A.T. Sustainable renewable energy resources utilization in rural areas. *Renew. Sustain. Energy Rev.* **2016**, *66*, 1–9. [CrossRef]
5. Mandelli, S.; Barbieri, J.; Mereu, R.; Colombo, E. Off-grid systems for rural electrification in developing countries: Definitions, classification and a comprehensive literature review. *Renew. Sustain. Energy Rev.* **2016**, *58*, 1621–1646. [CrossRef]
6. Luthander, R.; Widén, J.; Nilsson, D.; Palm, J. Photovoltaic self-consumption in buildings: A review. *Appl. Energy* **2015**, *142*, 80–94. [CrossRef]

7. Evans, A.; Strezov, V.; Evans, T.J. Assessment of utility energy storage options for increased renewable energy penetration. Renew. *Sustain. Energy Rev.* **2012**, *16*, 4141–4147. [CrossRef]
8. Ministerio para la Transición Ecológica. *Real Decreto 244/2019, de 5 de Abril, Por el Que se Regulan Las Condiciones Administrativas, Técnicas y Económicas del Autoconsumo de Energía Eléctrica*; Ministerio para la Transición Ecológica: Madrid, Spain, 2018.
9. Stoppato, A.; Cavazzini, G.; Ardizzon, G.; Rossetti, A. A PSO (particle swarm optimization)-based model for the optimal management of a small PV(Photovoltaic)-pump hydro energy storage in a rural dry area. *Energy* **2014**, *76*, 168–174. [CrossRef]
10. Topriska, E.; Kolokotroni, M.; Dehouche, Z.; Wilson, E. Solar hydrogen system for cooking applications: Experimental and numerical study. *Renew. Energy* **2015**, *83*, 717–728. [CrossRef]
11. Bahramirad, S.; Reder, W.; Khodaei, A. Reliability-Constrained Optimal Sizing of Energy Storage System in a Microgrid. *IEEE Trans. Smart Grid* **2012**, *3*, 2056–2062. [CrossRef]
12. Pan, M.-A.; Lau, T.-K.; Tang, Y.; Wu, Y.-C.; Liu, T.; Li, K.; Chen, M.-C.; Lu, X.; Ma, W.; Zhan, C. 16.7%-efficiency ternary blended organic photovoltaic cells with PCBM as the acceptor additive to increase the open-circuit voltage and phase purity. *J. Mater. Chem. A* **2019**, *7*, 20713–20722. [CrossRef]
13. Fisac, M.; Villasevil, F.X.; López, A.M. High-efficiency photovoltaic technology including thermoelectric generation. *J. Power Sources* **2014**, *252*, 264–269. [CrossRef]
14. Zerhouni, F.Z.; Zerhouni, M.H.; Zegrar, M.; Benmessaoud, M.T.; Stambouli, A.B.; Midoun, A. Proposed methods to increase the output efficiency of a photovoltaic (PV) system. *Acta Polytech. Hung.* **2010**, *7*, 55–70.
15. Roldan-Blay, C.; Escrivá-Escrivá, G.; Roldán-Porta, C.; Álvarez-Bel, C. An optimisation algorithm for distributed energy resources management in micro-scale energy hubs. *Energy* **2017**, *132*, 126–135. [CrossRef]
16. Roldán-Blay, C.; Escrivá-Escrivá, G.; Roldán-Porta, C. Improving the benefits of demand response participation in facilities with distributed energy resources. *Energy* **2018**, *169*, 710–718. [CrossRef]
17. Ismail, M.S.; Moghavvemi, M.; Mahlia, T.M.I. Techno-economic analysis of an optimized photovoltaic and diesel generator hybrid power system for remote houses in a tropical climate. *Energy Convers. Manag.* **2013**, *69*, 163–173. [CrossRef]
18. Maheri, A. Multi-objective design optimisation of standalone hybrid wind-PV-diesel systems under uncertainties. *Renew. Energy* **2014**, *66*, 650–661. [CrossRef]
19. Goel, S.; Sharma, R. Performance evaluation of stand alone, grid connected and hybrid renewable energy systems for rural application: A comparative review. *Renew. Sustain. Energy Rev.* **2017**, *78*, 1378–1389. [CrossRef]
20. Ghafoor, A.; Munir, A. Design and economics analysis of an off-grid PV system for household electrification. *Renew. Sustain. Energy Rev.* **2015**, *42*, 496–502. [CrossRef]
21. Maleki, A.; Hajinezhad, A.; Rosen, M.A. Modeling and optimal design of an off-grid hybrid system for electricity generation using various biodiesel fuels: A case study for Davarzan, Iran. *Biofuels* **2016**, *7*, 669–712. [CrossRef]
22. Shin, H.; Hur, J. Optimal Energy Storage Sizing With Battery Augmentation for Renewable-Plus-Storage Power Plants. *IEEE Access* **2020**, *8*, 187730–187743. [CrossRef]
23. Twaha, S.; Ramli, M.A. A review of optimization approaches for hybrid distributed energy generation systems: Off-grid and grid-connected systems. *Sustain. Cities Soc.* **2018**, *41*, 320–331. [CrossRef]
24. Zhou, W.; Lou, C.; Li, Z.; Lu, L.; Yang, H. Current status of research on optimum sizing of stand-alone hybrid solar–wind power generation systems. *Appl. Energy* **2010**, *87*, 380–389. [CrossRef]
25. Cîrstea, S.D.; Moldovan-Teselios, C.; Cîrstea, A.; Turcu, A.C.; Darab, C.P. Evaluating Renewable Energy Sustainability by Composite Index. *Sustainability* **2018**, *10*, 811. [CrossRef]
26. Kung, C.-C.; McCarl, B.A. Sustainable Energy Development under Climate Change. *Sustainability* **2018**, *10*, 3269. [CrossRef]
27. Ntanos, S.; Skordoulis, M.; Kyriakopoulos, G.; Arabatzis, G.; Chalikias, M.; Galatsidas, S.; Batzios, A.; Katsarou, A. Renewable Energy and Economic Growth: Evidence from European Countries. *Sustainability* **2018**, *10*, 2626. [CrossRef]
28. Armeanu, D.Ş.; Vintilă, G.; Gherghina, Ş.C. Does Renewable Energy Drive Sustainable Economic Growth? Multivariate Panel Data Evidence for EU-28 Countries. *Energies* **2017**, *10*, 381. [CrossRef]
29. Bilan, Y.; Streimikiene, D.; Vasylieva, T.; Lyulyov, O.; Pimonenko, T.; Pavlyk, A. Linking between Renewable Energy, CO2 Emissions, and Economic Growth: Challenges for Candidates and Potential Candidates for the EU Membership. *Sustainability* **2019**, *11*, 1528. [CrossRef]
30. Zsiborács, H.; Baranyai, N.H.; Vincze, A.; Zentkó, L.; Birkner, Z.; Máté, K.; Pintér, G. Intermittent Renewable Energy Sources: The Role of Energy Storage in the European Power System of 2040. *Electronics* **2019**, *8*, 729. [CrossRef]
31. Cebotari, S.; Benedek, J. Renewable Energy Project as a Source of Innovation in Rural Communities: Lessons from the Periphery. *Sustainability* **2017**, *9*, 509. [CrossRef]
32. Ali, S.; Jang, C.-M. Optimum Design of Hybrid Renewable Energy System for Sustainable Energy Supply to a Remote Island. *Sustainability* **2020**, *12*, 1280. [CrossRef]
33. Reddy, P.D.P.; Reddy, V.C.V.; Manohar, T.G. Ant Lion optimization algorithm for optimal sizing of renewable energy resources for loss reduction in distribution systems. *J. Electr. Syst. Inf. Technol.* **2018**, *5*, 663–680.
34. Bashir, M.; Sadeh, J. Size optimization of new hybrid stand-alone renewable energy system considering a reliability index. In Proceedings of the 2012 11th International Conference on Environment and Electrical Engineering, Venice, Italy, 18–25 May 2012; pp. 989–994.

35. Kim, S.-G.; Jung, J.-Y.; Sim, M.K. A Two-Step Approach to Solar Power Generation Prediction Based on Weather Data Using Machine Learning. *Sustainability* **2019**, *11*, 1501. [CrossRef]
36. Hammad, E. Resilient cooperative microgrid networks. *IEEE Trans. Ind. Inform.* **2019**, *16*, 1539–1548. [CrossRef]
37. Wallsgrove, R.; Woo, J.; Lee, J.-H.; Akiba, L. The Emerging Potential of Microgrids in the Transition to 100% Renewable Energy Systems. *Energies* **2021**, *14*, 1687. [CrossRef]
38. Quiles, E.; Roldán-Blay, C.; Escrivá-Escrivá, G.; Roldán-Porta, C. Accurate Sizing of Residential Stand-Alone Photovoltaic Systems Considering System Reliability. *Sustainability* **2020**, *12*, 1274. [CrossRef]
39. Nosratabadi, S.M.; Hooshmand, R.-A.; Gholipour, E. A comprehensive review on microgrid and virtual power plant concepts employed for distributed energy resources scheduling in power systems. *Renew. Sustain. Energy Rev.* **2017**, *67*, 341–363. [CrossRef]
40. Billinton, R.; Jonnavithula, A. Application of sequential monte carlo simulation to evaluation of distributions of composite system indices. *IEEE Proc. Gener. Transm. Distrib.* **1997**, *144*, 87–90. [CrossRef]
41. Billinton, R.; Allan, R.N. *Reliability Evaluation of Power Systems*; Springer: Boston, MA, USA, 1984.

applied sciences

MDPI

Article

Long-Term Forecasting Potential of Photo-Voltaic Electricity Generation and Demand Using R

Karina Vink [1,2,3,*], Eriko Ankyu [1,4] and Yasunori Kikuchi [1,5]

1 Technology Integration Unit, Global Research Center for Environment and Energy based on Nanomaterials Science (GREEN), National Institute for Materials Science (NIMS), Tsukuba, Ibaraki 305-0044, Japan (during study); ankyu.eriko.gn@u.tsukuba.ac.jp (E.A.); ykikuchi@ifi.u-tokyo.ac.jp (Y.K.)
2 Department of Water Engineering and Management (WEM) Faculty of Engineering Technology, University Twente, 7522 NB Enschede, the Netherlands
3 Department of Construction Management and Engineering (CME), Faculty of Engineering Technology, University Twente, 7522 NB Enschede, the Netherlands
4 Faculty of Human Sciences, University of Tsukuba, Tsukuba, Ibaraki 305-8577, Japan
5 Institute for Future Initiatives, the University of Tokyo, Tokyo 113-8654, Japan
* Correspondence: k.vink@utwente.nl; Tel.: +31-53-489-2547

Received: 18 May 2020; Accepted: 23 June 2020; Published: 28 June 2020

Abstract: For micro-grid cost-benefit analyses, both energy production and demand must be estimated on the long-term of one year. However, there remains a scarcity of studies predicting energy production and demand simultaneously and in the long-term. By means of programming in R and applying linear, non-linear, and support vector regression, we show the in depth analysis of the data of a micro-grid on solar power generation and building energy demand and its potential to be modeled simultaneously on the term of one year, in relation to electricity costs. We found solar power generation is linearly related to solar irradiance, but the effect of temperature on total output was less pronounced than anticipated. Building energy demand was found to be related to multiple parameters of both time and weather, and could be estimated through a quadratic function in relation to temperature. Models for both solar power generation and building energy demand could predict electricity costs within 8% of actual costs, which is not yet the ideal accuracy, but shows potential for future studies. These results provide important statistics for future studies where building energy consumption of any building type is correlated in detail to various time and weather parameters.

Keywords: energy demand; long-term forecasting; machine learning; R programming; solar power generation; support vector regression

1. Introduction

Increasing the share of renewable energy is crucial to overcome climate change, and to target 7.2 of the sustainable development goals to be reached by 2030. Micro-grids are increasingly becoming more popular as solutions for helping to reach renewable energy goals [1]. Micro-grids allow for local energy production, often combined with local energy storage, and often combined with an awareness or ambition to reduce overall energy demand. As micro-grids involve the management of both energy production and demand, both of these sides need to be predicted simultaneously in order to generate future cost benefit analyses. What we have found is that there is still a wide gap between predicting energy production and demand, in terms of scale, time, methodology, and software.

In this study we aim to determine the potential of long-term forecasting (at least one year) of energy production and demand on the scale of a research building by means of programming with the software R and applying linear, non-linear, and support vector regression. The research building in question is the National Institute of Materials Science (NIMS) NanoGREEN/WPI-MANA building in Tsukuba, Japan, which has a micro-grid consisting of four arrays of solar panels and has been in operation since 2013. The raw data per second of this micro-grid contained gaps and erroneous data, which have been cleansed in a previous study as described [2]. Using the improved data records from 2015 to 2017, we evaluated the potential of predicting future energy demand and energy generation.

Published studies on optimizing building energy consumption have increased 10 fold in the past 10 years [3]. This has included numerous variations of machine learning studies, which have explored either the forecasting of building energy demand [4], or renewable energy generation and weather patterns (for solar irradiance) [5], separate from each other. However, for micro-grid costs estimations it is necessary to consider these two factors simultaneously. A second shortcoming of previous studies is that they tend to focus on short term predictions and neglect long-term predictions. For micro-grid cost and benefit analyses we require long-term future predictions of the amount of solar power generated and the energy demand of the building. What is considered long- and short-term is defined differently for energy demand and weather prediction, as seen in Appendix A. According to the sources in Appendix A, the long-term for solar irradiance, our main weather parameter of interest, ranges from one day to two months, whereas for energy demand this is 1-10 years. Ideally it would be useful to predict the costs and benefits of the micro-grid one year ahead, with monthly averages and hourly estimates, as the electricity price varies per hour. We expect to see reduced solar power generation over time as the efficiency of the solar panels decreases, and an increase in building energy demand due to higher occupancy and laboratory use over the years. So far no study has considered how to incorporate both energy demand and solar power generation by means of machine learning for the long-term of one year. To reach this, it is important to consider not just the optimum results from higher-level regression analysis, but also the underlying physical and chemical mechanisms causing the found results in the actual data and suggest practical solutions based on these principles.

The hypotheses for solar power generation are the following. Larson et al. [6] showed a linear relationship between solar power generation and solar irradiance. However, it is also shown by Dupré et al. [7] that temperatures higher than 25 degrees Celsius negatively affect the capability of crystalline silicon to generate electricity. Similarly, it should perform between 1 and 1.1 times the normalized efficiency between 25 and 0 degrees Celsius, and presumably even high at colder temperatures. It is therefore expected to see an influence of temperature as well. Degradation is expected to lead to reduced efficiency in 2017 as compared to 2015.

The hypotheses for building energy demand are the following: the total building energy demand consists of purchased electricity, solar power generation, and battery discharging. The amount of generated solar power is at most around 10% of the total building energy demand and is immediately consumed by the building [8]. To accurately link back to electricity prices, only purchased electricity is considered as it is the main contributor. It is expected that building energy demand depends on building occupancy as well as weather circumstances, with colder and warmer temperatures leading to higher use of air-conditioning.

2. Materials and Methods

2.1. Data Description

An overview of the parameter names and units is given in Table 1.

- Micro-grid data, two parameters. An initial cost analysis was performed on the hourly data from the micro-grid by [8] from April 2013 to March 2018. These data were recorded and obtained from a different data storage point in the Energy Management System and not yet cleansed. The data used in this study concerns per second data, from January 2015 to December 2017, which was obtained from the micro-grid data storage point and was cleansed as described in [9]. It was converted into hourly data to enable analysis with hourly weather data by means of averaging. Parameters of interest are purchased electricity and total solar power generation.

For the purpose of this study, battery charging and discharging is ignored as it is of negligible proportions. The electricity amount and cost reduction function of the battery is active only during July, August, and September. Previous studies showed this amount is maximally 4.86% of the total building energy demand [10].

- Japan Meteorology Agency (JMA) data, six parameters. The hourly weather parameters for the corresponding time period of January 2015 to December 2017 were downloaded in csv files from the JMA website (freely accessible) [11] in batches of one month due to size restrictions. Parameters of interest are temperature, precipitation, sunlight, wind speed, solar irradiation, and humidity.
- Time parameters (6): year (note that due to the total data volume each year is calculated separately), month, day of month, day of week, hour, holiday.
- Electricity cost data, one parameter (see Figure 1). These data were obtained from the Facility Management Office of NIMS, based on the yearly renewed contract with the Tokyo Electric Power Company. The Feed in Tariff surcharge for renewable energy is included. Other factors such as fuel regulatory costs, basic connection to the electricity grid and spare line fees are not included since these are shared by the entire site and independent of building electricity use.

Table 1. Parameter names and units.

Data Source	Parameter Name	Parameter Unit
NanoGREEN/WPI-MANA micro-grid	total_pv_ac purchased_electricity_kwh	kWh (alternating current) kWh
Japan Meteorological Agency (JMA)	temperature precipitation sunlight wind_speed solar_irradiation humidity	°C mm numeral (between 0–1 h) m/s W/m^2 %
Time parameters	month day_of_month hour_end day_of_week holiday	month (1–12) day (1–31) hour (1–24) day (1–7, starting on Monday) numeral (0 for no holiday, 1 for holiday)
National Institute of Materials Science (NIMS) Facility Management Office	electricity price	JPY (Japanese Yen)

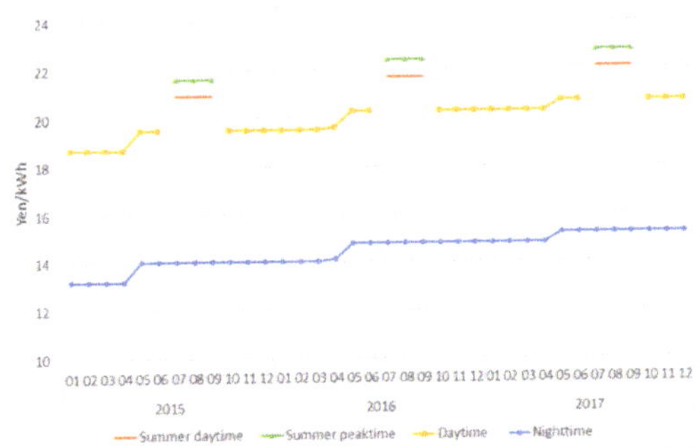

Figure 1. Electricity costs per time period (Japanese Yen/kWh).

2.2. Approach

Machine learning is the ability of a computer to learn from data without explicit programming. The machine learns from the existing data and predicts future data. In supervised learning, algorithms build a mathematical model of a set of data that contains both the inputs and the desired outputs, so their classification is known [12]. Supervised machine learning can be applied to the data in two forms:

- Predictive modeling to determine what would be the output if a new (future) input is applied (to estimate future demand and solar power generation)
- Prescriptive modeling to optimization what actions should be taken given the data (to estimate the impact of additional storage devices)

In this study we want to examine the potential of predicting two parameters: solar power generation and building energy use. We endeavor to evaluate the prediction potential of both parameters with one model using the same input parameters for weather and time, to see to which extent accurate electricity cost forecasts can be achieved.

2.3. Motivation for Software Choice

Various software is available for performing machine learning. Given the overview of advantages and disadvantages shown in Appendix B, we chose the software R (R version 4.0.1) mainly as it deals efficiently with large datasets, is freely available, and has high data analysis and visualization potential.

2.4. Data Cleansing

After the initial cleansing and averaging of the per second micro-grid data as per [9], additional cleansing steps were required. The JMA weather data similarly required cleansing. The cleansing of both datasets are described below. In all cases of missing data, the entire row of data was removed so as not to lead to invalid correlations.

1. Removing missing values from the micro-grid data and replacing negative values with 0. For the micro-grid, this concerned the maintenance days in November (2015: November 14, 8:00–20:00, 2016: November 19, 8:00–19:00, 2017: November 18, 8:00–18:00). Two additional instances were found to have missing values (2015/11/13 12:00, 2017/09/25 21:00). For solar power generation, 269 negative values were recorded during nighttime in 2017, which were replaced with 0.
2. Removing JMA data with poor data quality. The JMA has excellent record keeping concerning data quality. Most values have the data quality value of "8: There are no missing data on the statistical basis (normal value)". The number of instances when the data quality was less than this are described in Table 2. Values with data quality values of 4 (10 instances) or 5 (2 instances) were kept. There were several instances where data was missing for multiple parameters at the same time.
3. Removing outliers (see SM1: R code for outlier removal). Note that the parameter (relative) humidity is in % unit, and sunlight is relative to one full hour (1), and therefore these parameters should not have outliers removed. The hourly electricity price is given based on predetermined conditions and cannot have outliers. Precipitation occurs intermittently and in irregular amounts, meaning it is prone to outliers. This leaves three weather parameters and the two micro-grid parameters to remove outliers for. For the parameters "solar irradiation" and "total pv ac" outliers were calculated without taking the values of 0 into account. Extreme outliers were removed by calculating three times the interquartile range (iqr) using the following code:

$$\text{extreme\_threshold\_upper} = (\text{iqr} * 3) + \text{upperq} \tag{1}$$

$$\text{extreme\_threshold\_lower} = \text{lowerq} - (\text{iqr} * 3) \tag{2}$$

Boxplot results were used to show the difference before and after outlier removal. As an example, Figure 2 shows the data before and after removing outliers for the parameter wind speed, using data from 2015. Outliers were checked for overlap between parameters and consequently removed. Before outlier removal the standard deviation, mean, minimum and maximum values were calculated for the five parameters: wind speed, temperature, solar irradiation, total pv ac, and purchased electricity, for each year. Table 3 gives an overview of the found outliers.

4. Removing additional outliers occurring around maintenance days. After removal of the values in steps 1–3, we observed several remaining uncharacteristic data values occurring around the maintenance days. As example, from November 13 to 16 in 2015 the values for purchased electricity from the micro-grid were uncharacteristically low. The cause of these low values is most likely the shutdown of electricity dependent equipment in anticipation of and in reaction to the maintenance procedures, during which all electricity through the main power line is stopped. This led to the removal of 47 rows of data in 2015; 47 in 2016 and 49 in 2017. After this final removal step the standard deviation, mean, minimum and maximum values were calculated for the five parameters wind speed, temperature, solar irradiation, total pv ac, and purchased electricity, for each year.

Table 2. Overview of data quality values of JMA parameters.

	Year	2015	2016	2017	2015	2016	2017	2015	2016	2017
#	JMA weather parameter data quality	5: There is a deficit of 20% or less of the data that is the basis of statistics (quasi-normal value)			4: There is a deficit exceeding 20% in the data underlying the statistics (lack of data)			1: There is no statistical value (missing)		
1	Temperature	-	-	-	-	-	-	-	-	2
2	Precipitation	-	-	-	-	-	-	-	-	-
3	Sunlight	-	-	1	-	1	3	-	-	22
4	Wind speed	-	-	-	-	-	-	-	1	2
5	Solar irradiance	1	-	-	2	2	2	1	-	2
6	Humidity	-	-	-	-	-	-	-	-	2

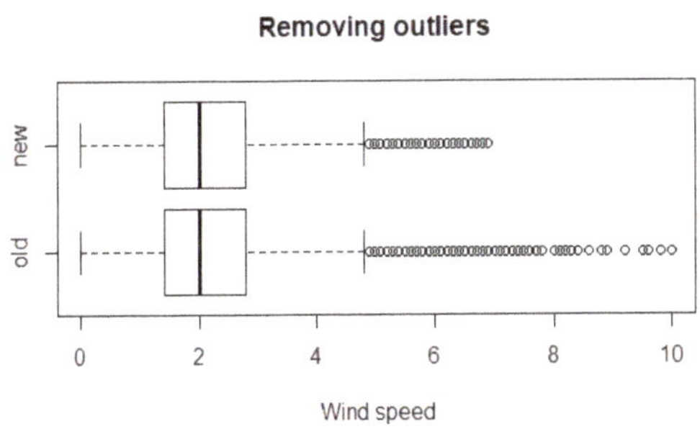

Figure 2. Data points before and after removing outliers of the parameter "Wind speed" in data from 2015.

Table 3. Outliers removed per parameter and per year.

Parameter Name	2015	2016	2017
wind_speed	88	68	100
temperature	0	0	0
solar_irradiation	0	0	0
total_pv_ac	0	0	0
purchased_electricity_kwh	2	0	0

2.5. Data Exploration (See SM2: R Code for Data Exploration Graphs)

The main exploration of the relationships between the different parameters was performed through means of plotting different combinations of parameters:

- Generated solar power/building energy demand in relation to various individual parameters per month and hour of the day, with values color coded by e.g., temperature, based on Boeye [13].
- All parameters correlated to each other through the ggpairs package, based on [14].

These results show that removing outliers of wind speed, total pv ac and purchased electricity does not greatly affect the standard deviation of all five parameters. It is interesting to note that whereas the maximum of solar irradiation has increased from 2015 to 2016 and stayed nearly the same between 2016 and 2017, the maximum solar power generated (total pv ac) is decreasing every year.

3.2. Data Exploration

The top of Figure 4 shows the amount of generated solar power per month and hour of the day, with values color coded by temperature; the bottom shows the amount of building energy demand for the same criteria; both for the year 2015. The top shows the expected relation of more solar power being generated during hours of sunlight. The bottom shows a clear baseline of energy demand related to storage and other research related equipment in the building, with peaks in energy demand in winter and summer months for the heating, ventilation, and air-conditioning system.

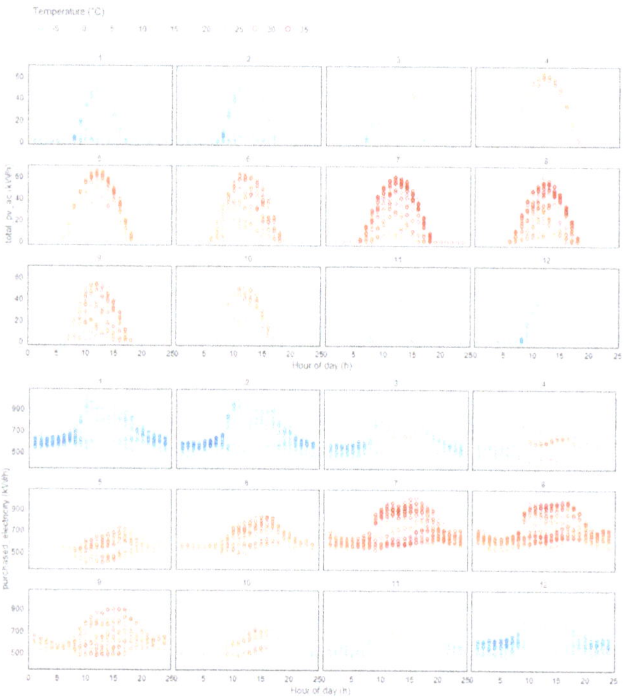

Figure 4. Generated solar power/building energy demand per month and hour of the day, with values color coded by temperature; 2015 data.

Figure 5 shows a cross comparison between 13 tested factors for correlation, with the data from 2015. For all years, solar power generation and solar irradiance were found to be highly correlated to each other (0.988–0.991), and both were found to be highly correlated to sunlight, and inversely correlated to humidity. They were both weakly correlated to temperature and wind speed. Other weather factors were also correlated in various degrees to each other and to the time parameters month and hour. Purchased electricity (building energy demand without micro-grid contributions) was found to be highly correlated to the day of the week, hour of the day, and temperature extremes (high from July to September, and especially in the morning hours of December to February). It is weakly correlated to holidays and other weather parameters sunlight, wind speed, solar irradiation, and humidity (inversed).

Figure 5. All parameters correlated to each other through the ggpairs package; 2015 data.

3.3. Solar Power Generation

The linear model had the format of (y = a + b * x). Smoothing the curve through means of support vector regression (red line, formed by x symbols, in Figure 6) shows a better fit than through the linear model (LM) (blue line, formed by x symbols, in Figure 6). Table 6 shows the difference in Root Square Mean Errors (RSME) for each year, proving the worth of applying support vector regression for 2015 and 2016, though it takes some computing time. For 2017, surprisingly the linear regression model showed less error than the SVR. This means that SVR cannot be applied haphazardly but must be tested against linear regression models. The values for epsilon and cost in Table 6 and Figure 7 show the epsilon parameters were all low, leaning towards overfitting of the models. The cost parameter found in 2016 was rather high, which also creates allows for few errors during classification. As SVR generates new models randomly, it may be necessary to compare the results of several runs repetitively even though one execution already creates dozens of models, in order to examine the differences.

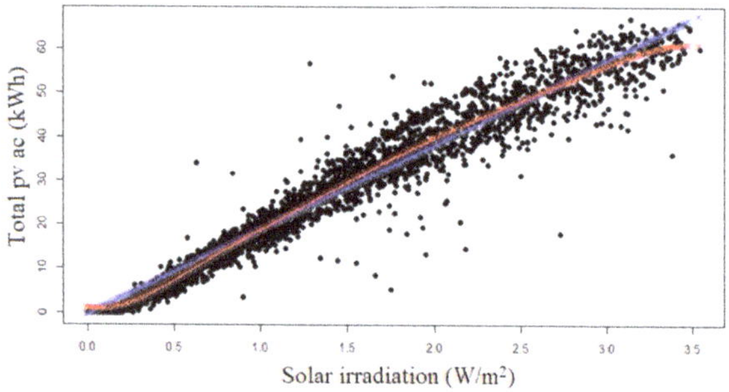

Figure 6. Linear (blue line) and Support Vector (red line) Regression of two parameters for 2015 data.

Table 6. Coefficients and Root Square Mean Errors (RSME) and parameters of Linear Models (LM) (y = a + b * x), Support Vector Regression (SVR), and tuned SVR models.

Year	Intercept (a)	Slope (b)	LM RSME	SVR RSME	Tuned SVR RSME	Epsilon	Cost
2015	−0.5101	19.2985	2.527038	2.37958	2.263242	0.08	8
2016	−0.5088	18.4918	2.204646	1.974913	1.888035	0.06	128
2017	−0.4922	18.0549	2.181584	2.37958	2.263037	0.08	4
Average	−0.5037	18.61506667					

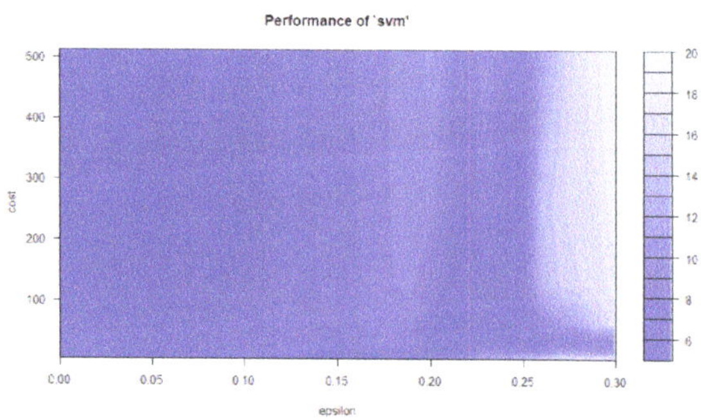

Figure 7. Epsilon and costs factors by SVR; 2015 data.

3.4. Building Energy Demand

Figure 8 shows the equation found through non-linear regression to best describe the relationship between purchased energy and temperature for 2015. This equation has the format of $y = a - b * x + c * x^2$, where y is purchased electricity, x is temperature, a is the intercept, b is the slope, and c the exponential curve. Results from all years are shown in Table 7, including the error difference between linear and non-linear models. The non-linear model error is lower than the linear model error for all cases. However, the data seem to show two superimposed u-shaped curves, suggesting a third parameter is affecting the results and it may be possible to generate two equations to better model the data. To enable this, the data were separated into 8 cases as described in Table 4. The subsequent results showed no clear effect of any combination of parameters on the purchased electricity. Figure 9 shows the effect of temperature, month, and hour on the purchased electricity from 2015. This illustrates there is no distinguishable repetitive difference in effects between temperature and hours of the day during which the building is occupied on the purchased electricity, which was expected.

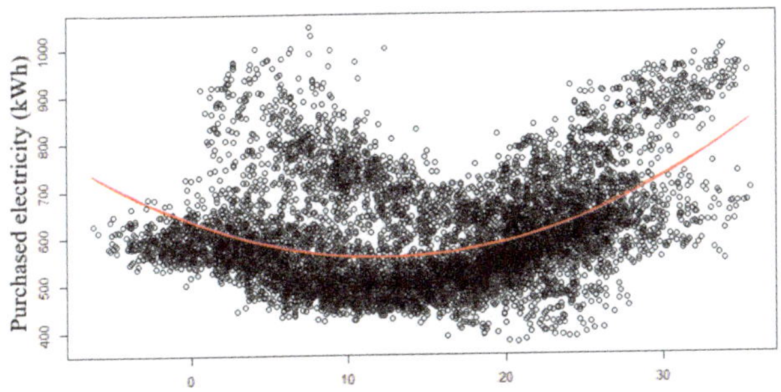

Figure 8. Quadratic equation relating purchased electricity to temperature; 2015 data.

Table 7. Parameters for the quadratic equation (6) describing the relationship between temperature and purchased electricity. All had 1 iteration to convergence.

Year	LM Error	NLM Error	Intercept (a)	Slope (b)	Curve (c)	Residual Sum of Squares	Tolerance
2015	107.3204	99.04455	633.1747	12.7784	0.5251	84,442,961	1.83×10^{-9}
2016	113.8353	103.4177	660.1028	13.3542	0.6305	92,577,853	4.283×10^{-9}
2017	109.8714	102.973	698.3077	9.8411	0.4722	90,956,240	8.543×10^{-9}
Average			663.8617333	11.99123333	0.5426		

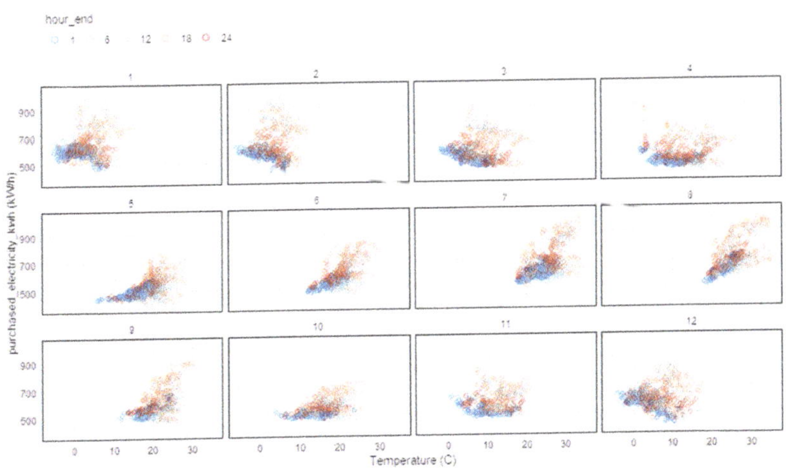

Figure 9. Purchased electricity per month, temperature, and hour of the day; 2015 data.

3.5. Electricity Prices

Table 8 shows the results of applying the predicted equations for solar power generation and purchased electricity to the electricity price totals in comparison to the actual electricity price totals. The equations' accuracy varied from +6 to −8% of actual prices, with purchased energy showing the greatest variation.

Table 8. Electricity prices in 1000 JPY and % correctly modeled per micro-grid parameter (costs saved vs. remaining costs) for both actual data and estimated data based on modeled equations.

Parameter	2015	2015	%	2016	2016	%	2017	2017	%
total_pv_ac (costs saved)	1579	1557	−1.36%	1594	1643	+3.08%	1651	1740	+5.41%
purchased_electricity_kwh (costs made)	86,669	92,049	+6.21%	99,253	97,304	−1.96%	107,646	99,142	-7.90%

4. Discussion

Two different types of models were used to estimate the relationships between the micro-grid parameters on the one hand, and weather and time parameters on the other. This is due to the dependencies found in the exploratory data analysis phase.

This analysis showed a linear relationship between solar power generation and solar irradiance, which was not always improved by applying support vector regression. The effects of temperature on solar power generation were different than anticipated. Figure 4 shows that whereas higher temperatures of 30–35 degrees Celsius indeed lead to less produced solar power (a maximum of around 60 kWh whereas colder temperatures showed higher production); lower temperatures of −5-0 degrees Celsius lead to far less produced solar power of 40 kWh. Although the efficiency may be higher, the lack of solar irradiation has a far greater effect than temperature. The electricity costs saved by generated solar power were overestimated by 3% in 2016 and 5% in 2017, suggesting that degradation has a considerable effect on the total generated solar power and, although hidden, should be taken into account in future models.

To examine the hypothesis that temperature could have a large influence on purchased energy, a quadratic equation was found to describe this relationship. At times when the building was occupied, temperature extremes did show a higher effect on purchased electricity (Figure 4). Applying the equation to hourly electricity costs led to over- and underestimations of the actual costs of 6% and 8% respectively. With annual costs running up to around 100,000,000 JPY, it is desirable to have an accuracy within 1% or less, which this simplified equation has not yet achieved. Purchased energy was found to be dependent on numerous other variables, namely, month, hour of day, day of week, holiday or not, and other weather parameters. Splitting the data into groups as per table IV did not lead to a clear distinction, as the month of the year still had a substantial influence as shown in Figure 9. A more inclusive approach incorporating all of these parameters might produce better results, especially if the hidden factor of building occupancy, or of purchased electricity base demand, is estimated for each hour.

To enable predictions of up to one year into the future by means of machine learning, the first step in improving these equations are to incorporate solar power generation efficiency degradation on the one hand, and the complex interactions of time parameters leading to building occupancy on the other hand. If these improvements can lead to more accurate cost predictions within 1% error margin, the second step is to predict associated weather parameters for up to one year, mainly solar irradiance and temperature. As these two parameters influence each other, and this study showed that purchased electricity had a higher correlation to solar irradiance than to temperature (Figure 5), it may even be possible to have a single weather parameter predict both micro-grid parameters. This would simplify the number of unknowns that have to be predicted. Alternatively, if many weather factors can be predicted with great accuracy, the reverse approach may be taken. Sharma et al. [5] found that solar irradiance can be inferred from the parameters day of the year, temperature, dew point, wind speed, sky cover (alternatively sun hours), precipitation, and humidity. If the input weather parameters can be predicted with sufficient accuracy, it is worthwhile to compare the results to using only solar irradiance as input.

Although linear, non-linear, and support vector regression were adequate in uncovering the relationships and the strength of relationship between the various parameters, it was not yet successful in defining the complex interplay that leads to purchased electricity. A future study applying principal component regression, such as done by Davo et al. [19], may resolve how to best model purchased electricity with the greatest accuracy combined with the least number of parameters.

5. Conclusions

The equations resulting from linear, non-linear, and support vector regression allowed for long-term (one year) predictions of related electricity costs that were within 7.9% accuracy of actual costs. This study further found evidence supporting the potential to model both solar power generation and building energy consumption simultaneously, based on the parameter of solar irradiance and in combination with known time parameters affecting building occupancy.

This study is the first to attempt to predict both energy demand and production on a long-term scale using R software in combination with actual micro-grid data records. The cross correlations generated between all input parameters and generated equations provide important indications as to how to integrate energy demand and production into one model with potentially one weather parameter, solar irradiance, as input. For future studies on building energy consumption and solar power generation, this study provides an important basis for evaluating which weather and time parameters to consider during data exploration, as well as correlation methods. It furthermore shows how applying actual data records can assist directly in improving model accuracy. Finally, all the documentation on R codes used during the data exploration and analysis phases is provided as appendices and Supplementary Materials, which facilitates replicating the steps taken with other datasets.

Supplementary Materials: The following are available online at http://www.mdpi.com/2076-3417/10/13/4462/s1, SM1: R code for outlier removal, SM2: R code for data exploration graphs, SM3: R code for linear regression and SVR, SM4: R code for plotting nonlinear regression; SM5: R code for data rows graph. Datasets related to this article can be found at https://doi.org/10.6084/m9.figshare.c.4176698, hosted at figshare (2018), or https://doi.org/10.11503/nims.1003, hosted at NIMS IMEJI (2018).

Author Contributions: Conceptualization, K.V. and Y.K.; methodology, K.V.; software, K.V.; validation, K.V.; formal analysis, K.V. and E.A.; investigation, K.V. and E.A.; data curation, E.A. and K.V.; writing—original draft preparation, K.V.; writing—review and editing, K.V., E.A., and Y.K.; visualization, K.V.; supervision, Y.K.; project administration, K.V. All authors have read and agreed to the published version of the manuscript.

Funding: Part of the research is financially supported by the MEXT Program for Integrated Materials Development.

Conflicts of Interest: The authors declare no conflict of interest.

Notes on References: In order to encourage gender equality in STEM research, the authors have included a minimum of 50% references of which the first author appears to be female. For more information about this goal, see: Collins, K. Conversation with Dr. Emily M. Gray. Journal and Proceedings of the Gender Awareness in Language Education Special Interest Group 2019, 11, pp. 71-80, SSN 1884-152X. Feb. 2019. URL: https://gale-sig.org/wp-content/uploads/2019/04/Winter-18-19-Full-Issue-v4.pdf.

Appendix A

Table A1. Differing forecasting horizons from previous prediction studies.

Parameter	Long Term	Medium Term	Short Term	Reference
Solar irradiance	1–3 days	1–6 h	15 min–2 h	[20]
Solar irradiance	28–45 h (1 day)	1–6 h	-	[21]
Solar irradiance	1 month	1 day	1 min	[22]
Building energy demand	1–10 years	1 month–1 year	1 h–1 week	[23]
Building energy demand	> 1 year	1 week–1 year	1 h–1 week	[24]

Appendix B

Table A2. Advantages and disadvantages of different software for supervised machine learning [25,26].

Software	R	Python	Java	SAS	SPSS
Description	Developed by statisticians, aimed at handling numerical values; procedural language (step by step subroutines)	General purpose programming language, simple to read, used for websites; object oriented (bundles procedures and data together as parts of objects)	Programming language and computer platform; object oriented (bundles procedures and data together as parts of objects)	Statistical Analysis System, developed for analyzing large quantities of agriculture data	Statistical Package for the Social Sciences, the first statistical programming language for the PC
Advantages	Free, easy to learn, many options for data visualization, suited for statistical analysis and explanatory and predictive modeling	Free, easy to learn, built-in debugging	High speed	Easy to handle large datasets	Good user interface, official support, easy to learn and write code, similar to Excel
Disadvantages	No official support or user interface	Slow with large datasets	Not strong in statistical modeling and visualization	High learning curve	Slow with large datasets

Appendix C

Excel formula for electricity prices:
IF(OR(K3 = 1,I3 = "Sunday"),VLOOKUP(C3,electricityprice!C$2:H$74,5,FALSE),
IF(AND(J3<>"Summer",P3 > = 8,P3 < = 21),VLOOKUP(C3,electricityprice!C$2:H$74,4,FALSE),
IF(AND(J3<>"Summer",P3<=7),VLOOKUP(C3,electricityprice!C$2:H$74,5,FALSE),
IF(AND(J3<>"Summer",P3>=22),VLOOKUP(C3,electricityprice!C$2:H$74,5,FALSE),
IF(AND(J3 = "Summer",P3 > = 13,P3 < = 15),VLOOKUP(C3,electricityprice!C$2:H$74,3,FALSE),
IF(AND(J3 = "Summer",P3 > = 16,P3 < = 21),VLOOKUP(C3,electricityprice!C$2:H$74,2,FALSE),
IF(AND(J3 = "Summer",P3 > = 8,P3 < = 12),VLOOKUP(C3,electricityprice!C$2:H$74,2,FALSE),
IF(AND(J3 = "Summer",P3 < = 7),VLOOKUP(C3,electricityprice!C$2:H$744,5,FALSE),
IF(AND(J3 = "Summer",P3 > = 22),VLOOKUP(C3,electricityprice!C$2:H$74,5,FALSE),
-999999)))))))))

References

1. González, R.M.; Wattjes, F.D.; Gibescu, M.; Vermeiden, W.; Slootweg, J.G.; Kling, W.L. Applied Internet of Things Architecture to Unlock the Value of Smart Microgrids. *IEEE Internet Things J.* **2018**, *5*, 5326–5336. [CrossRef]
2. Vink, K.; Ankyu, E.; Koyama, M. Multiyear microgrid data from a research building in Tsukuba, Japan. *Sci. Data* **2019**, *6*, 190020. [CrossRef] [PubMed]
3. Mocanu, E. Machine Learning Applied to Smart Grids. Ph.D. Thesis, Eindhoven University of Technology, Eindhoven, the Netherlands, 2017.
4. Mocanu, E.; Nguyen, P.H.; Gibescu, M.; Kling, W.L. Comparison of machine learning methods for estimating energy consumption in buildings. In Proceedings of the 2014 International Conference on Probabilistic Methods Applied to Power Systems (PMAPS), Durham, UK, 7–10 July 2014.
5. Sharma, N.; Sharma, P.; Irwin, D.; Shenoy, P. Predicting solar generation from weather forecasts using machine learning. In Proceedings of the 2011 IEEE International Conf. on Smart Grid Communications (SmartGridComm), Brussels, Belgium, 17–20 October 2011; pp. 528–533.
6. Larson, D.P.; Nonnenmacher, L.; Coimbra, C.F.M. Day-ahead forecasting of solar power output from photovoltaic plants. *Renew. Energ.* **2016**, *91*, 11–20. [CrossRef]
7. Dupré, O.; Vaillon, R.; Green, M.A. Thermal Issues in Photovoltaics and Existing Solutions. In *Thermal Behavior of Photovoltaic Devices*, 1st ed.; Dupré, O., Vaillon, R., Green, M.A., Eds.; Springer: Berlin/Heidelberg, Germany, 2017; pp. 1–28. [CrossRef]

8. Vink, K.; Ankyu, E.; Elborg, M.; Koyama, M. Performance and cost analysis of building scale micro-grid operation. *Energ. Procedia* **2019**, *156*, 425–429. [CrossRef]
9. Ankyu, E.; Vink, K.; Koyama, M. Economic evaluation of big data analysis and peak cut operation of micro grid system (Japanese). In Proceedings of the Japan Computer Chemistry Society 2018 Fall meeting, Hirosaki, Japan, 3–4 November 2018. [CrossRef]
10. Ankyu, E.; Vink, K.; Koyama, M. Optimization of the micro-grid system using the actual data of the NanoGREEN/WPI-MANA building. In Proceedings of the 20th GREEN Symposium, Tsukuba, Japan, 19 February 2019.
11. Japan Meteorological Agency. Past Weather Data Download (Japanese). 2018. Available online: http://www.data.jma.go.jp/gmd/risk/obsdl/index.php (accessed on 18 May 2020).
12. Russell, S.J.; Norvig, P. Supervised Learning. In *Artificial Intelligence: A Modern Approach*, 3rd ed.; Russell, S.J., Norvig, P., Eds.; Prentice Hall: Upper Saddle River, NJ, USA, 2010; pp. 695–697.
13. Solar Panel Analysis Part 1: Exploration. Available online: https://www.jeroenboeye.com/blog/solar-panel-analysis-pt-1-exploration/ (accessed on 18 May 2020).
14. Using Linear Regression for Predictive Modeling in R. Available online: https://www.dataquest.io/blog/statistical-learning-for-predictive-modeling-r/ (accessed on 18 May 2020).
15. Support Vector Regression with R. Available online: https://www.svm-tutorial.com/2014/10/support-vector-regression-r/ (accessed on 18 May 2020).
16. Karatzoglou, A.; Meyer, D. Support Vector Machines in R. *J. Stat. Softw.* **2006**, *15*. [CrossRef]
17. Performing Nonlinear Least Square and Nonlinear Regressions in R. Dataconomy. Available online: https://dataconomy.com/2017/08/nonlinear-least-square-nonlinear-regression-r/ (accessed on 18 May 2020).
18. BBC Visual and Data Journalism Cookbook for R Graphics, How to Create BBC Style Graphics. Available online: https://bbc.github.io/rcookbook/ (accessed on 18 May 2020).
19. Davo, F.; Alessandrini, S.; Sperati, S.; Delle Monache, L.; Airoldi, D.; Vespucci, M.T. Post-processing techniques and principal component analysis for regional wind power and solar irradiance forecasting. *Solar Energ.* **2016**, *134*, 327–338. [CrossRef]
20. Diagne, H.M.; David, M.; Lauret, P.; Boland, J.; Schmutz, N. Review of solar irradiance forecasting methods and a proposition for small-scale insular grids. *Renew. Sustain. Energ. Rev.* **2013**, *27*, 65–76. [CrossRef]
21. Persson, C.; Bacher, P.; Shiga, T.; Madsen, H. Multi-site solar power forecasting using gradient boosted regression trees. *Solar Energ.* **2017**, *150*, 423–436. [CrossRef]
22. Voyant, C.; Notton, G.; Kalogirou, S.; Nivet, M.L.; Paoli, C.; Motte, F.; Fouilloy, A. Machine learning methods for solar radiation forecasting: A review. *Renew. Energ.* **2017**, *105*, 569–582. [CrossRef]
23. Raza, M.Q.; Khosravi, A. A review on artificial intelligence based load demand forecasting techniques for smart grid and buildings. *Renew. Sustain. Energ. Rev.* **2015**, *50*, 1352–1372. [CrossRef]
24. Mocanu, E.; Nguyen, H.P.; Gibescu, M.; Kling, W.L. Deep learning for estimating building energy consumption. *Sustain. Energ. Grids Netw.* **2016**, *6*, 91–99. [CrossRef]
25. Ozgur, C.; Colliau, T.; Rogers, G.; Hughes, Z. MatLab vs. Python vs. R. *J. Data Sci.* **2016**, *15*, 355–372. [CrossRef]
26. Java, P.R. Scala: Many Languages that Big Data Speaks. Available online: https://www.linkedin.com/pulse/java-python-r-scala-many-languages-big-data-speaks-debajani/ (accessed on 18 May 2020).

applied sciences

MDPI

Article

Enhanced Quasi Type-1 PLL-Based Multi-Functional Control of Single-Phase Dynamic Voltage Restorer

Hafiz Ahmed [1], Samet Biricik [2], Hasan Komurcugil [3] and Mohamed Benbouzid [4,5,*]

1 Nuclear Futures Institute, Bangor University, Bangor LL57 2DG, UK; hafiz.h.ahmed@ieee.org
2 Department of Electrical and Electronic Engineering, European University of Lefke, Lefke 99728, Turkey; s.biricik@eul.edu.tr
3 Department of Computer Engineering, Eastern Mediterranean University, Mersin 10, Famagusta 99628, Turkey; hasan.komurcugil@emu.edu.tr
4 Institut de Recherche Dupuy de Lôme (UMR CNRS 6027 IRDL), University of Brest, 29238 Brest, France
5 Logistics Engineering College, Shanghai Maritime University, Shanghai 201306, China
* Correspondence: mohamed.benbouzid@univ-brest.fr; Tel.: +33-2-98018007

Abstract: This paper considers the reference signal generation problem for the multi-functional operation of single-phase dynamic voltage restorers. For this purpose, a single-phase quasi type-1 phase-locked loop (QT1-PLL) is proposed. The pre-loop filter part of this PLL is composed of a frequency-fixed delayed signal cancellation method and a two-stage all-pass filter. Thanks to the frequency-fixed nature, the pre-loop filter is easy to implement and can provide rejection of any measurement offset. Moreover, this PLL benefits from the excellent harmonic robustness property of the conventional QT1-PLL. Small-signal modeling and gain tuning procedures are detailed in this paper. In order to track the reference voltage signals generated by the proposed PLL, a super-twisting sliding mode controller is also presented, which helps to achieve fast dynamic responses. Laboratory-scale prototype-based experimental studies were conducted to validate the developed reference generator and the controller. Experimental results show that the proposed method is fast in detecting and compensating any grid voltage anomalies to maintain constant load voltage despite voltage sag, swell, and harmonic distortions.

Keywords: grid-synchronization; dynamic voltage restorer; converter control system; sliding mode control

Citation: Ahmed, H.; Biricik, S.; Komurcugil, H.; Benbouzid, M. Enhanced Quasi Type-1 PLL-Based Multi-Functional Control of Single-Phase Dynamic Voltage Restorer. *Appl. Sci.* **2022**, *12*, 146. https://doi.org/10.3390/app12010146

Academic Editor: Edris Pouresmaeil

Received: 5 November 2021
Accepted: 21 December 2021
Published: 24 December 2021

1. Introduction

Power quality problems such as voltage sag, swell, and harmonics can significantly affect the performance of critical loads that are used in hospital, water treatment plant, data center, etc. In addition to harmonics, voltage sag and swells are quite common in power grid and can be caused by many reasons such as lightning strike, accident, short circuit, over loading, switching on or off large electrical loads, etc. In the case of water treatment plant, voltage sag/swell can interrupt the treatment process including dissolved air flotation, filtration, and disinfection. Any interruption in the process can take up to 8 h to resolve, causing one-third production loss for the day. This highlights the importance of mitigating voltage-related power quality problems.

In order to address power quality issues such as voltage sag, swell, and harmonics, the dynamic voltage restorer (DVR) became very popular in recent times [1–4]. DVR can compensate voltage sag, swell, and harmonics to maintain desired constant voltage at the critical load side. In the literature, various topologies of DVR are proposed. However, the most basic DVR topology is made of any dc voltage source such as battery, photovoltaic panel, etc., together with a voltage source inverter with LC filter and a transformer. The transformer provides galvanic isolation between the inverter and its secondary is connected between the grid voltage and the load in series. The control system of DVR constantly monitors grid voltage and injects compensation voltage when it detects any deviation from the desired voltage. This results in maintaining desired load voltage despite various power quality disturbances at the grid.

The detection of voltage sag, swell, and harmonics plays a vital role in ensuring the effective operation of DVR. Voltage sag and swells are typically short-lived incidents with a time range of few milliseconds to a minute [5]. As such, fast and accurate detection of sag and swell is essential for fast responsive DVR operation. In the ideal case, voltage sag/swell can be detected very quickly by comparing the voltage magnitude with respect to the ideal magnitude. However, calculating the magnitude can be tricky in the presence of nonlinearities such as harmonics.

The first step in compensation voltage calculation is to generate the reference voltage, which should be in-phase with the measured grid voltage. If the grid voltage is ideal, i.e., has a frequency of 50 Hz (or 60 Hz) and contains no harmonics, then generating the reference voltage is straightforward. However, this is not the case in practice. According to the European standard EN 50160 [6], grid voltage can vary between -3 Hz and $+2$ Hz of the nominal frequency. However, the grid frequency has to be within 1% of the nominal value, i.e., ± 0.5 Hz for 99.5% of the time. Moreover, harmonics are almost always present in the grid due to the ever increasing penetration of nonlinear loads and converter-interfaced distributed energy sources. All these factors complicate the reference voltage calculation.

In order to address the non-ideal characteristics of the measured grid voltage, researchers often rely on phase-locked loop (PLL) [7–15] or similar techniques to generate the reference voltage for DVR. Using PLL, first, the instantaneous phase of the grid voltage fundamental component is estimated. This can then be used as the unit template for the reference voltage. By multiplying the unit template with the desired amplitude, the actual reference grid voltage can be calculated. In this study, we are considering a single-phase DVR. However, traditional PLLs in the form of synchronous reference frame (SRF) PLLs work only for three-phase systems. For single-phase systems, an additional orthogonal signal generator (OSG) is required to implement SRF-PLL. In the single-phase DVR literature, various OSGs have already been used. In [16], a second-order generalized integrator (SOGI) is used as the OSG. The effect of DC offset is not considered in [16]. Moreover, despite having band-pass filtering property, traditional SOGI-PLL cannot completely reject the dominant harmonics. Thus, multiple SOGIs need to be placed in parallel in order to reduce the effect of dominant harmonic components. This can make the overall system computationally complex. It is to be noted here that the conventional SRF-PLL can be made robust to harmonics and DC offset by considering a slow loop-filter, i.e., by reducing the bandwidth. This has been considered in [17]. Although this is an interesting practical solution, this strategy is suitable for voltage compensation but not harmonics.

Self-tuning filter (STF) is similar to SOGI; however, in the case of STF, the error feedback is independent of the grid frequency. In [18,19], the authors have applied STF as the reference generator without considering frequency adaptation. As such, the reference generator in [18,19] can be limiting when the grid frequency varies significantly. Quarter-delay is another popular method for generating orthogonal signal. This approach is considered in [20]. However, in off-nominal frequency condition, the required delay would be fractional which increases computational complexity. In [21], an adaptive notch filter (ANF) is used as the reference generator. However, the considered ANF did not use gain normalization. This can make the convergence time slow in the presence of voltage sag.

Based on the literature review, it is clear that there is demand for a reference generator that is robust to harmonics and DC offset while at the same time is computationally simple to implement. For this purpose, in this study, a quasi type-1 PLL [22–24] is considered. This PLL has been selected for several reasons. Firstly, it can provide amplitude normalization without using any low-pass filter unlike conventional SRF-PLL, cf. [25]. Use of additional low-pass filter in the amplitude normalization block will introduce tuning complexity as there is one more gain to tune. Secondly, this PLL provides good harmonic robustness thanks to the use of a moving-average filter. Finally, it has only one gain to tune unlike SRF-PLL where the loop-filter has two tuning parameters. However, this PLL can work only for three-phase system. Thus, a single-phase version of this PLL is proposed in this study.

For this purpose, first, a half-cycle delayed signal cancellation method is applied to reject the DC offset. Then, a frequency-fixed all-pass filter (APF) [26] is applied to generate the orthogonal signal. However, single-stage APFs will generate double frequency oscillations in off-nominal frequency conditions. As such, a two-stage APF is considered that can eliminate the double frequency error. The APF is selected in this study as it can generate orthogonal signals without using any tuning gain unlike other choices available in the literature such as SOGI [16,27], ANF [21], etc. This is beneficial from the tuning simplicity point of view. Finally, the filtered grid voltage signal and its orthogonal component are used as the inputs to the quasi type-1 PLL. A small-signal model of the proposed PLL is developed and validated. Finally, tuning of the PLL gain is also presented. Compared to conventional PLL techniques summarized in [28], our approach is very simple to tune as it has only one tuning gain. All the techniques summarized in [28] have at least three parameters to tune if amplitude normalization is considered. Unlike most of the techniques in [28], our quadrature signal generator is frequency non-adaptive. As such, there is no frequency feedback which may be beneficial from the stability point of view.

Once the reference signal is generated, the role of control system is to follow/track the reference. In the literature, synchronous frame approach in the form of proportional-integral (PI) controllers [29] are widely used. Although this controller can be easily designed and implemented, the dynamic response can be slow. In order to enhance dynamic performance, advanced controllers such as model predictive control (MPC) [7,8], $\mathcal{H}_\infty$ [30], and sliding mode control (SMC) [31–33] are proposed in the literature. MPC can be sensitive to model parameters mismatch. Computational complexity can be an issue for $\mathcal{H}_\infty$ controllers. SMC is often a suitable choice for controlling nonlinear systems in the presence of parameter mismatch and/or external perturbations. As such, this technique has been selected in this study.

The rest of this article is organized as follows: Section 2 explains the operation of the proposed DVR together with the error model for controller development. Development of the proposed reference signal generator is provided in detail in Section 3. Development of the sliding-mode controller is provided in Section 4. Experimental results on a laboratory-scale prototype together with simulation results are provided in Section 5. Finally, Section 6 concludes this paper.

2. DVR Modeling and Problem Formulation

In this study, we consider a single-phase DVR. A connection diagram of the DVR is given in Figure 1. In this configuration, the DVR, which is a full-bridge voltage source inverter, is connected in series through a transformer to the protected load. Isolation between the load the DVR is provided by the transformer. Grid voltage sensor is used to continuously monitor the deviation from the reference voltage, and the appropriate compensation voltage is injected by the DVR to maintain the ideal voltage at the protected load terminal. Filter current and compensation voltage dynamics of the DVR are given by the following:

$$\frac{di_f}{dt} = \frac{1}{L_f}(v_i - v_c), \quad (1)$$

$$\frac{dv_c}{dt} = \frac{1}{C_f}\left(i_f - i_g\right), \quad (2)$$

where v_i, v_g, v_c, i_f, i_g, L_f, and C_f denote the input voltage by the DVR, grid voltage, compensation voltage, filter current, filter inductance, and filter capacitance, respectively. The DVR voltage can be expressed as $v_i = uV_{dc}$ where the DC-link voltage is denoted by V_{dc}, and the control signal is given by u. Moreover, in the ideal case, one can also write the following.

$$v_c = v_g - v_L. \quad (3)$$

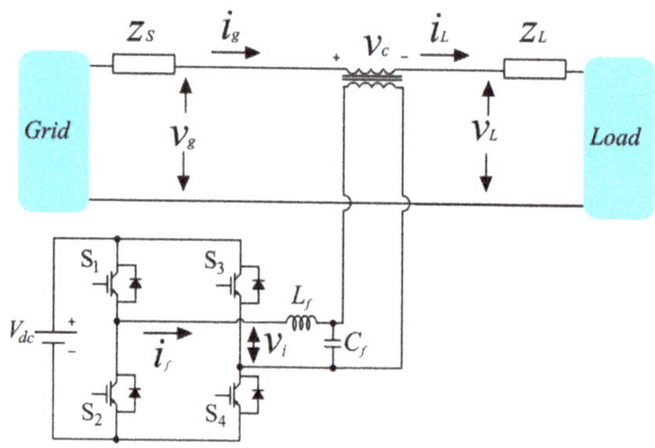

Figure 1. Topology of the considered DVR.

In normal operating condition, the compensation voltage given by (3) will be zero. However, in practice, the grid voltage is never ideal. Various grid anomalies such as voltage sag/swell, harmonics, noise, etc., are present. In this case, (3) can no longer be used for the compensation voltage calculation. In order to ensure that the sensitive load voltage remains as close as possible to the desired voltage, reference voltage needs to be calculated and this voltage should be in-phase with the grid voltage fundamental for the efficient operation by the DVR. Let us consider that the reference voltage is denoted by v_c^{ref}. Then, the tracking error and its derivative are given by the following.

$$\xi_1 = v_c^{\text{ref}} - v_c, \quad (4)$$

$$\xi_2 = \dot{\xi}_1 = \dot{v}_c^{\text{ref}} - \dot{v}_c. \quad (5)$$

Then, by substituting the DVR dynamical Equations (1) and (2) into (4) and (5), the DVR tracking error dynamics can be obtained as follows:

$$\frac{d\xi_1}{dt} = \xi_2, \quad (6)$$

$$\frac{d\xi_2}{dt} = -\delta\xi_1 + \delta u V_{dc} + w, \quad (7)$$

where the coefficient $\delta = 1/\left(L_f C_f\right)$ and the perturbation term $w(t)$ is given by the following.

$$w(t) = -\frac{1}{C_f}\frac{di_g}{dt} - \delta v_c^{\text{ref}} - \ddot{v}_c^{\text{ref}}.$$

It is assumed that the perturbation term $w(t)$ has a bounded derivative and is upper bounded by $\mid \frac{dw}{dt} \mid \leq W, W > 0$. Thus, the control of DVR is essentially divided into two parts. In part 1, the problem is to generate the reference voltage v_c^{ref} from the measured grid voltage v_g. In part 2, the problem is to find the control signal u that will ensure that the tracking errors converge asymptotically to zero. These two issues are addressed in the following two sections.

3. Reference Signal Generator

In this Section, reference voltage v_c^{ref} will be generated from the measured grid voltage v_g. For this purpose, let us consider the singe-phase grid voltage signal model in time-domain as provided below:

$$v_g(t) = v_0 + \underbrace{V_p \sin(\theta_g)}_{v_g^{\varnothing}}, \quad (8)$$

where $|v_0| \geq 0$, V_p, and $\theta_g = \int \omega_g dt, \theta_g(0) = \phi$ are the DC offset, amplitude, and instantaneous phase with ω_g being the grid frequency, and ϕ is the initial phase angle. The frequency is $\omega_g = \omega_n + \delta_\omega$, where $\omega_n = 100\pi$ is the nominal frequency and δ_ω is the deviation from the nominal frequency. For efficient operation of the DVR, the reference voltage should be in phase with the instantaneous phase of the grid voltage θ_g. Thus, the process of extracting θ_g from the measured voltage v_g is considered in this section. For this purpose, a PLL-based approach has been considered in this study. Details are provided below.

3.1. DC Offset Rejection

Measurement offset v_0 causes estimation error in the estimated phase. Thus, rejection of this offset is essential in order to eliminate the steady-state error. For this purpose, the half-cycle delayed signal cancellation (DSC) method is a popular choice in the literature [34,35]. The same approach is considered here. For this purpose, let us consider half-cycle delayed version of the signal v_g as follows:

$$\begin{aligned} v_g^{t_d}(t) &= v_0 + V_p \sin(\omega_g(t - t_d) + \phi), \\ &= v_0 - V_p \sin(\theta_g), \end{aligned} \tag{9}$$

where $t_d = \frac{T}{2}$ with $T = \frac{2\pi}{\omega_g}$ being the period of the grid voltage signal. Then, the DC offset can be eliminated by the following operation.

$$\hat{v}_g^{\varnothing} = \frac{1}{2}\left(v_g - v_g^{t_d}\right). \tag{10}$$

In implementing Equation (10), the actual period of the grid voltage is required. In the off-nominal frequency condition, the amount of required dely could be a fraction, thereby increasing computational complexity. A potential solution is to use the nominal period; however, this will introduce amplitude and phase attenuation in the off-nominal condition. In this study, only extracting the phase is required. thus, appropriate compensation of the phase delay is necessary to eliminate the error. For this purpose, let us consider the transfer function of the DSC operation (10) as provided by the following:

$$\frac{\hat{v}_g^{\varnothing}}{v_g}(s) = G_{dc}(s) = \frac{1 - e^{-t_d s}}{2}, \tag{11}$$

where the estimated value is indicated by $\hat{}$. The discrete-time version of the transfer function (11) is given by the following:

$$G_{dc}(z) = \frac{1 - z^{-N_d}}{2}, \tag{12}$$

where the required delay is given by as $N_d = t_d f_s$ with f_s being the sampling frequency. By substituting $s = j\omega_g$, the phase angle of the transfer function (11) is given by the following.

$$\begin{aligned} \angle G_{dc}(s) &= \tan^{-1}\left\{\tan\left(\frac{\pi}{2} - \frac{T\omega_g}{4}\right)\right\}, \\ &= \frac{\pi}{2} - \frac{T\omega_g}{4}, \\ &= \frac{\pi}{2} - \frac{T}{4}(\omega_n + \delta_\omega), \\ &= \frac{\pi}{2} - \frac{T}{4}\frac{2\pi}{T} - \frac{T}{4}\delta_\omega, \\ &= -\underbrace{\frac{T}{4}}_{k_{dc}}\delta_\omega. \end{aligned} \tag{13}$$

In calculating (13), it is assumed that $\omega_n = 2\pi/T$. Equation (13) shows that the use of nominal period in (10) causes a phase delay of $-k_{dc}\delta_\omega$ in the off-nominal frequency case. This phase needs to be compensated in the loop-filter to eliminate the phase error in the off-nominal frequency condition.

3.2. Tuning-Free Fixed-Frequency Orthogonal Signal Generator

Once the DC offset is eliminated, the signal $v_g^{\emptyset}$ can be used to generate an orthogonal signal component. In this study, we are considering an all-pass filter (APF) [26,36]. APF is a first-order filter, and it can be used to generate orthogonal signals without any tuning gain. This filter can be used either as frequency-adaptive or non-adaptive configurations. In this study, a frequency-fixed operation is considered same as the DC offset rejection method, as highlighted in Section 3.1. The frequency-fixed APF transfer function is given by the following:

$$\frac{v_g^{\emptyset\perp}}{v_g^{\emptyset}}(s) = \text{APF}(s) = \frac{\omega_n - s}{\omega_n + s}, \tag{14}$$

where superscript $\perp$ indicates orthogonal signals. A time-domain block diagram of APF is provided in Figure 2. Similarly to Section 3.1, the frequency-fixed operation of the APF will introduce amplitude and phase attenuation in the estimated orthogonal signal. As such, characterization of the APF is needed to determine the necessary compensation mechanism. Phase-angle of the transfer function (14) is given by the following.

$$\angle\text{APF}(s) = \tan^{-1}\left(\frac{2\omega_g\omega_n}{(\omega_g - \omega_n)(\omega_g + \omega_n)}\right). \tag{15}$$

For small frequency drift, i.e., $\delta_\omega \approx 0$, it can be assumed that $\omega_g + \omega_n \approx 2\omega_g$. Then, the phase angle (15) can be approximated as follows.

$$\begin{aligned}\angle\text{APF}(s) &\approx \tan^{-1}\left(\frac{1}{\frac{\omega_g}{\omega_n} - 1}\right). \\ &\approx -\frac{\pi}{2} - \tan^{-1}\left(\frac{\omega_g}{\omega_n} - 1\right), \\ &\approx -\frac{\pi}{2} - \left(-\frac{\pi}{4} + \frac{\frac{\omega_g}{\omega_n}}{2} + \frac{\left(\frac{\omega_g}{\omega_n}\right)^2}{2!} + \cdots\right).\end{aligned} \tag{16}$$

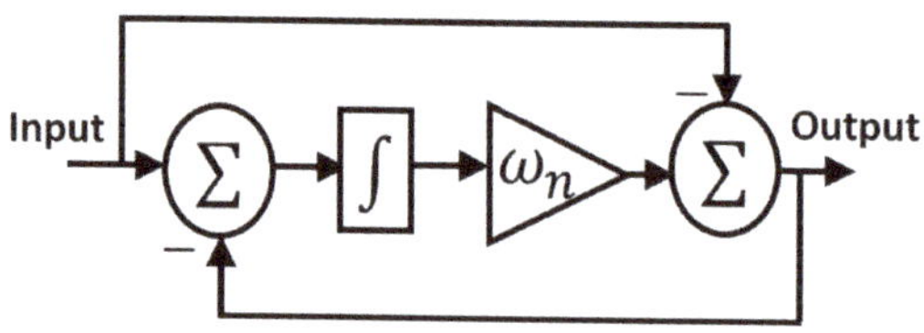

Figure 2. Time-domain implementation of frequency-fixed APF.

By ignoring the high-order (2nd and above) terms from the Taylor series expansion in (16), this equation can be simplified as follows.

$$\begin{aligned}\angle\mathrm{APF}(s) &\approx -\frac{\pi}{2} - \left(-\frac{\pi}{4} + \frac{\omega_g}{2\omega_n}\right),\\ &\approx -\frac{\pi}{4} - \frac{\omega_g}{2\omega_n},\\ &\approx -\frac{\pi}{4} - \frac{(\omega_n + \delta\omega)}{2\omega_n},\\ &\approx -\frac{\pi}{4} - \frac{1}{2} - \frac{\delta\omega}{2\omega_n}.\end{aligned} \tag{17}$$

As shown in (17), the first two terms are frequency independent while the third term depends on frequency variation. This term, i.e., $k'_\phi = 1/2\omega_n$, needs to compensated.

It is well known that single-phase grid voltage can be represented by an unbalanced two phase-system. It was shown in [36] that single-stage frequency-fixed APF cannot effectively remove the unbalanced component as the bandwidth of the unbalanced component rejection part is very narrow. As such, the unbalanced component will appear as double the fundamental frequency component after Park transformation. Thus, in order to eliminate this error, the double frequency component needs to be rejected. This issue can be solved by increasing the bandwidth of the notch component. In the literature, a two-stage APF has been suggested for this purpose. The two-stage APF effectively increases notch bandwidth and enables frequency-fixed APFs to reject off-nominal frequency unbalanced components. However, the two-stage APF will double the phase delay in the off-nominal frequency condition. As such, the required phase compensation value is computed by $k_\phi = 2k'_\phi = 1/\omega_n$.

3.3. Implementation in PLL

The previous two subsections detailed the procedure for obtaining DC offset eliminated signal $v_g^{\varnothing}$ and its orthogonal component $v_g^{\varnothing\perp}$. These signals can be used as the input to PLL. The overall block diagram of the proposed PLL is provided in Figure 3. This section details the operating principle of this PLL. Before describing the phase detector operation of this PLL in our case, let us consider APF-filtered signals in the steady-state:

$$v_\beta(t) = -V_p\cos(\theta_g + \delta_\phi), \tag{18}$$

$$v_g'^{\varnothing}(t) = -V_p\sin(\theta_g + 2\delta_\phi), \tag{19}$$

where δ_ϕ is the off-nominal frequency phase attenuation by each stage of the APF. As per Figure 3, using the signal $v_g^{\varnothing}(t)$ and $v'_g(t)$, $v_\alpha(t)$ is obtained as follows.

$$\begin{aligned}v_\alpha(t) &= \frac{v_g^{\varnothing}(t) - v_g'^{\varnothing}(t)}{2},\\ &\approx V_p\sin(\theta_g) + \delta_\phi\cos(\theta_g).\end{aligned} \tag{20}$$

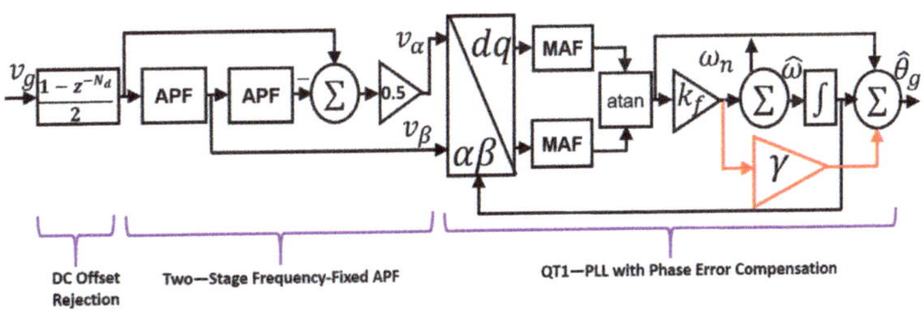

Figure 3. Overall block diagram of the proposed PLL.

The phase detector of the considered PLL is provided by the following.

$$\begin{bmatrix} v_d \\ v_q \end{bmatrix} = \begin{bmatrix} \cos(\hat{\omega}t) & \sin(\hat{\omega}t) \\ -\sin(\hat{\omega}t) & \cos(\hat{\omega}t) \end{bmatrix} \begin{bmatrix} v_\alpha \\ v_\beta \end{bmatrix}. \tag{21}$$

Then, direct-axis and quadrature-axis voltages can be obtained directly from (21). By applying a quasi-locked condition, i.e., $\omega - \hat{\omega} \approx 0$, the direct-axis and quadrature-axis voltages can be rewritten as follows.

$$\begin{aligned} v_d &= v_\alpha \cos(\hat{\omega}t) + v_\beta \sin(\hat{\omega}t), \\ &= \frac{V_p}{2}(\sin((\omega+\hat{\omega})t+\phi) + \sin((\omega-\hat{\omega})t+\phi)) + \frac{\delta_\phi}{2}(\cos((\omega-\hat{\omega})t+\phi)) \\ &\quad -\frac{V_p}{2}\big(\sin\big((\omega+\hat{\omega})t+\phi+2\delta_\phi\big) - \sin\big((\omega-\hat{\omega})t+\phi+2\delta_\phi\big)\big) \\ &\quad +\frac{V_p}{2}\frac{\delta_\phi}{2}(\cos((\omega+\hat{\omega})t+\phi)), \\ &\approx \frac{V_p}{2}\big(\sin(\phi)+\sin(\phi+\delta_\phi)\big) + \frac{V_p}{2}\big(\sin(2\hat{\omega}t+\phi) - \sin\big(2\hat{\omega}t+\phi+\delta_\phi\big)\big) \\ &\quad +\frac{\delta_\phi}{2}\cos(\phi) + \frac{\delta_\phi}{2}(\cos(2\hat{\omega}t+\phi)). \end{aligned} \tag{22}$$

$$\begin{aligned} v_q &= -v_\alpha \sin(\hat{\omega}t) + v_\beta \cos(\hat{\omega}t), \\ &= \frac{V_p}{2}(\cos((\omega+\hat{\omega})t+\phi) + \cos((\omega-\hat{\omega})t+\phi)) - \frac{\delta_\phi}{2}(\sin((\omega+\hat{\omega})t+\phi)) \\ &\quad -\frac{V_p}{2}\big(\cos\big((\omega+\hat{\omega})t+\phi+\delta_\phi\big) - \cos\big((\omega-\hat{\omega})t+\phi+\delta_\phi\big)\big) \\ &\quad +\frac{V_p}{2}\frac{\delta_\phi}{2}(\sin((\omega-\hat{\omega})t+\phi)), \\ &\approx \frac{V_p}{2}\big(\cos(\phi)+\cos(\phi+\delta_\phi)\big) + \frac{V_p}{2}\big(\cos(2\hat{\omega}t+\phi) - \cos\big(2\hat{\omega}t+\phi+\delta_\phi\big)\big) \\ &\quad +\frac{\delta_\phi}{2}\sin(\phi) + \frac{\delta_\phi}{2}\sin(2\hat{\omega}t+\phi). \end{aligned} \tag{23}$$

Double frequency components in (22) and (23) can easily be filtered by applying moving average filter (MAF) with half-cycle window length. The transfer functions of MAF in continuous and discrete domain are provided by the following.

$$\mathrm{MAF}(s) = \frac{1-e^{-t_d s}}{t_d s}, \tag{24}$$

$$\mathrm{MAF}(z) = \frac{1}{N}\frac{1-z^{-N_d}}{1-z^{-1}}. \tag{25}$$

By applying MAF to (22) and (23) and also assuming negligible off-nominal frequency phase shift, filtered v_d and v_q can be approximated as follows.

$$v'_d = V_p \sin(\phi), \tag{26}$$

$$v'_q = V_p \cos(\phi). \tag{27}$$

Filtered voltages are then fed to the loop-filter of QT1-PLL, as shown in Figure 3.

3.4. Small-Signal Modeling and Tuning

A small-signal model of the proposed PLL can be obtained by considering the signal flow in Figure 3. For this purpose, first, the small-signal model of the pre-loop filters needs to be developed. The first pre-loop filter is the delayed signal cancellation block, which is given by transfer function (11). This transfer function can be converted into synchronous reference frame by substituting $s = s + j\omega_n$.

$$\mathrm{DSC}_{dq}(s) = \frac{1 + e^{-t_d s}}{2}. \tag{28}$$

The transfer function of the APF in synchronous reference frame can be obtained by applying the Park transformation to single-stage APF and is given by the following Gautam et al. [26].

$$\mathrm{APF}_{dq}(s) = \frac{s + 2\omega_n}{2(s + \omega_n)}. \tag{29}$$

In our study, a two-stage APF is considered. As such, the effective transfer function is given by the following.

$$\mathrm{APF}^2_{dq}(s) = \left(\frac{s + 2\omega_n}{2(s + \omega_n)}\right)^2. \tag{30}$$

Considering the pre-loop filters, the small-signal model is shown in Figure 4, where $\gamma = k_{dc} + k_\phi$ is the overall phase compensation gain.

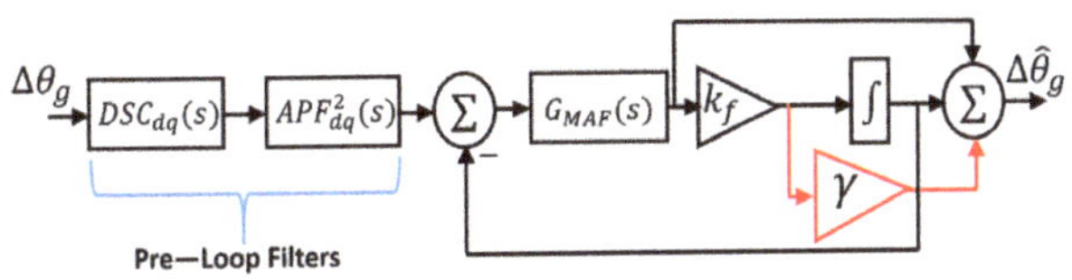

Figure 4. Small-signal model of the proposed PLL.

The proposed PLL has only one parameter to tune, which is the frequency estimation gain k_f. This gain can be tuned in several ways. Two of the popular approaches are based on open-loop phase margin and settling time. The later is considered in this study. In order to tune gain k_f using this method, a +2 Hz frequency step is considered. Considering a 2% settling time, settling time versus the gain k_f is given in Figure 5. From this figure, $k_f = 89$ has been found to provide the fastest settling time. This value has been considered as the optimal gain for k_f. Considering this value of k_f,validation of the small-signal model is provided in Figure 6.

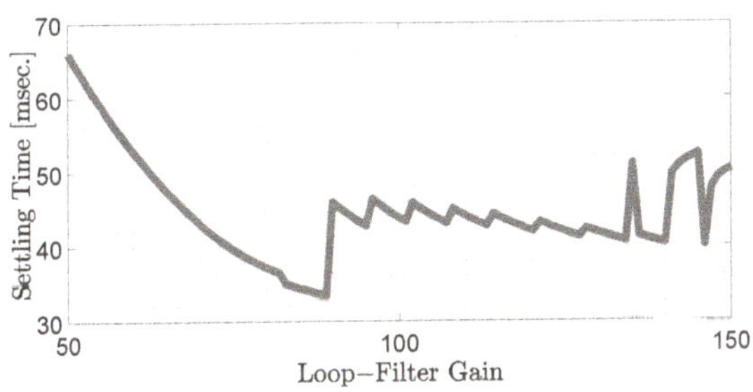

Figure 5. Settling time versus k_f for a +2 Hz frequency step change.

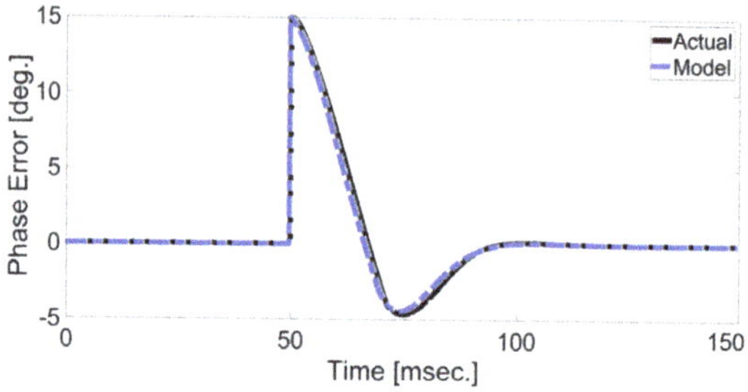

Figure 6. Small-signal model verification of the proposed PLL with $k_f = 89$.

Once the phase of the grid voltage fundamental component is estimated, the reference compensation voltage can be calculated as follows:

$$v_c^{\text{ref}} = V_p^{\text{ref}} \sin(\hat{\theta}_g), \tag{31}$$

where reference magnitude is provided by V_p^{ref}. The design of the control signal u based on the reference grid voltage (31) is detailed in the next section.

4. Super Twisting Sliding Mode Controller Design

In this Section, tracking error dynamics (6) and (7) will be used for the control design. For this purpose, let us consider that the control signal is composed of $u = u_0 + u_n$, where u_0 is the nominal control signal and u_n is the nonlinear part of the control signal. If we consider $u_0 = (1/V_{dc})\xi_1$ and $u_n = u_{ST}/(\delta V_{dc})$, then the total control signal can be written as follows.

$$u = \frac{1}{V_{dc}}\left(\xi_1 + \frac{u_{ST}}{\delta}\right). \tag{32}$$

Then, the tracking error dynamics (6) and (7) can be rewritten as follows.

$$\frac{d\xi_1}{dt} = \xi_2, \tag{33}$$

$$\frac{d\xi_2}{dt} = u_{ST} + w. \tag{34}$$

Tracking error dynamics (33) and (34) can be viewed as a perturbed second-order integrator. Numerous control techniques can be employed to stabilize the error under the presence of perturbation $w(t)$. In this study, we consider a second-order SMC [31,37] in the form of super-twisting SMC [38]. In order to design the super-twisting SMC, let us consider the following sliding surface.

$$\sigma_e = \xi_2 + \lambda_1\xi_1, \lambda_1 > 0. \tag{35}$$

Then, the controller u_{ST} in (34) can be designed as follows:

$$u_{ST} = -\lambda_1\xi_2 - \lambda_2|\sigma_e|^{\frac{1}{2}}\text{sgn}(\sigma_e) - \lambda_3\int_0^t \text{sgn}(\sigma_e(\tau))d\tau, \tag{36}$$

where sgn(.) is the conventional signum function and the gains λ_2 and λ_3 are selected as follows.

$$\lambda_3 > W, \lambda_2^2 > 4\lambda_3. \tag{37}$$

In order to analyze the stability of the controller (36), let us consider the derivative of the sliding surface (35) together with (33), (34), and (36) as provided below.

$$\dot{\sigma}_e = -\lambda_2|\sigma_e|^{\frac{1}{2}}\text{sgn}(\sigma_e) - \lambda_3\int_0^t \text{sgn}(\sigma_e(\tau))dt + w(t). \tag{38}$$

Let us consider the following variables.

$$\zeta_1 = \sigma_e, \tag{39}$$

$$\zeta_2 = -\lambda_3 \int_0^t \text{sgn}(\sigma_e(\tau))d\tau + w(t), \tag{40}$$

$$\frac{dw}{dt} = \eta(t). \tag{41}$$

Then, the dynamics of the sliding surface (38) can be rewritten as follows.

$$\frac{d\zeta_1}{dt} = -\lambda_2|\sigma_e|^{\frac{1}{2}}\text{sgn}(\sigma_e) + \zeta_2, \tag{42}$$

$$\frac{d\zeta_2}{dt} = -\lambda_3 \int_0^t \text{sgn}(\sigma_e(\tau))d\tau + \eta(t). \tag{43}$$

Then, for the selected control gain (37), finite-time convergence of the variables ζ_1 and ζ_2 can easily be established by using the results presented in Levant [37]. In order to implement the super-twisting controller, (32), (35), and (36) are required. An implementation block diagram of the super-twisting sliding-mode controller is provided in Figure 7.

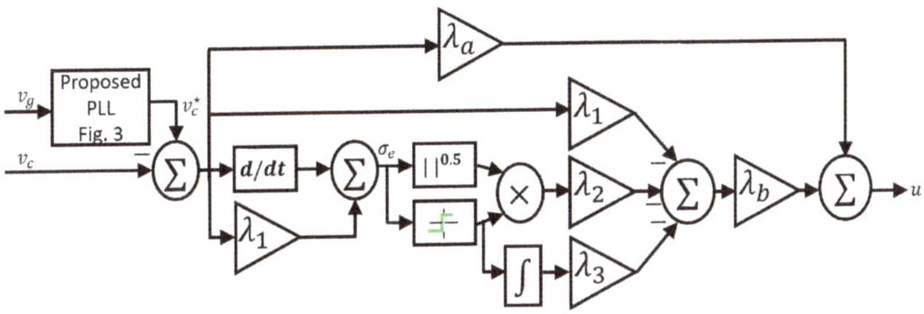

Figure 7. Block diagram of super-twisting sliding-mode controller with $\lambda_a = 1/V_{dc}$ and $\lambda_b = 1/(\delta V_{dc})$.

5. Simulation and Experimental Results

In this Section, simulation and experimental results are reported. The experimental setup used in this study is demonstrated in Figure 8. Here, a Texas Instrument C2000 series micro-controller is used to implement the proposed control and estimation algorithm. Parameters of the setup and control gains are provided in Table 1.

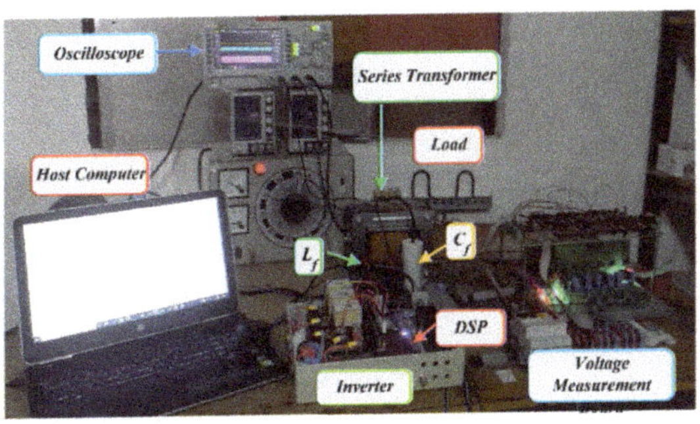

Figure 8. Experimental setup used in this study.

Table 1. Experimental study parameters.

System and Control Parameters	Values
Controller gains	
PLL gain	$k_f = 89$
Grid: voltage and frequency	120 V (rms) 50 Hz
Grid: impedance	$R_g = 1\ \mathrm{m}\Omega$, $L_g = 0.1\ \mu\mathrm{H}$
Transformer turns ratio	1:1
DC link voltage, V_{dc}	120 V
LC filter	$L_f = 0.8$ mH, $C_f = 50\ \mu$F
Sensitive series load	$R = 100\ \Omega$

Numerical simulation using Matlab/Simulink is conducted by considering the same values. In the first test, the grid voltage suddenly experiences a ≈30% sag. From approximately 170 V (peak), the grid voltage dropped to roughly 120 V (peak). Experimental results are provided in Figure 9. The results show that in order to mitigate the effect of grid voltage sag at the sensitive load side, the DVR was very quick to react and supplied the necessary 50 V in-phase compensation voltage. As a result, constant voltage was maintained at the load side. In the second test, voltage swell was considered, and the results are provided in Figure 10. Here, grid voltage increased to 210 V, which is roughly a 25% change from the nominal value. Unlike the first test, here, out-of-phase compensation voltage needs to be generated in order to reduce the voltage at the sensitive load end. As shown in Figure 10, the proposed enhanced QT1-PLL was very successful in generating the required roughly 40 V out-of-phase compensation voltage, and the sliding mode controller ensured the tracking of the reference compensation voltage by the DVR. These results show that the proposed approach can handle both sag and swell conditions. In the real grid, in addition to voltage sag/swell, harmonics are also a problem. The presence of harmonics will render the sensitive load voltage distorted. As a result, power quality degrades. In order to mitigate this issue, harmonic compensation is also required. In the final test, the grid voltage suddenly became distorted, and the results are provided in Figure 11. The proposed PLL can extract the fundamental component with high harmonic robustness. As a result, quick estimation of the grid harmonics was performed by PLL, and DVR injected the necessary harmonic compensation voltage to ensure very low distortion at the load side. The results in Figure 11 validate the suitability of the proposed PLL in a distorted grid condition.

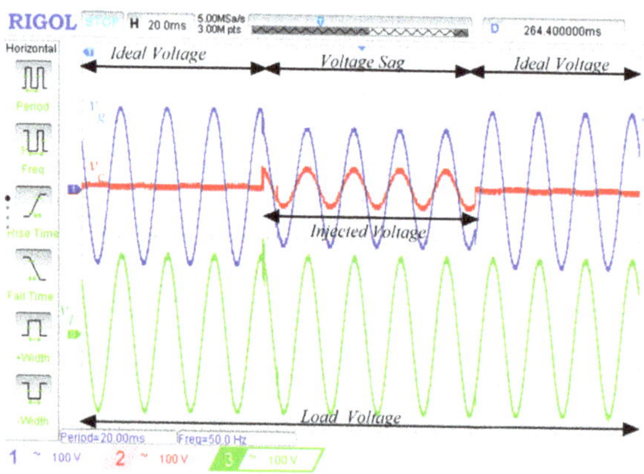

Figure 9. Experimental responses of v_g, v_c, and v_L subject to voltage sag in the grid.

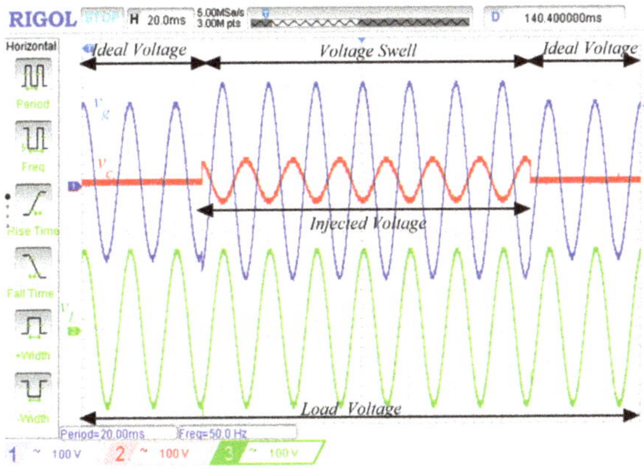

Figure 10. Experimental responses of v_g, v_c, and v_L subject to voltage swell in the grid.

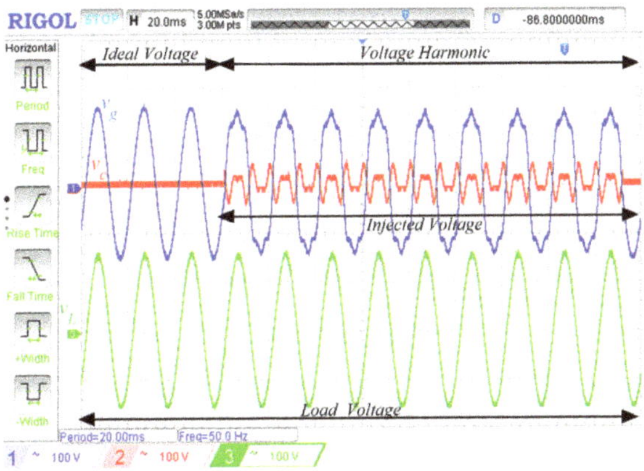

Figure 11. Experimental responses of v_g, v_c, and v_L subject to voltage harmonics in the grid.

In order to check the robustness of the used controller, we have used numerical simulations. In the simulation test, voltage sag and harmonics are considered. The nominal value of the filter inductor is 0.8 mH. This value is used to synthesize the control law. Simulation results with $\pm 25\%$ variation in the filter inductor are provided in Figure 12. The results in Figure 12 show that the response of DVR is very similar when the system parameter experienced a $\pm 25\%$ change from the nominal value. This shows that the sliding mode controller is robust to parameter variations.

Experimental results as shown in Figures 9–11 independently considered voltage sag, swell, and harmonics. In practice, in addition to these characteristics, phase and/or frequency of the grid voltage may also change simultaneously in the worst case scenario. In order to asses the performance of the proposed method, two additional simulation studies were considered. In the first test, grid voltage experienced -0.5 p.u. sag and -25° phase change simultaneously. In the second test, in addition to the sag, the grid voltage also experienced +25° phase change and +1 Hz frequency change. In both cases, the fault cleared after 100 ms and the grid became distorted after fault clearance. Numerical simulation results for the first and second test are provided in Figures 13 and 14. Results show that the proposed control method is very fast (roughly 1 cycle convergence time) despite very abrupt changes in grid voltage parameters. However, it is to be noted here that the load voltage transients are not that smooth compared to the case when only one parameter changed, such as in Figure 12. Smooth transient load voltage scan be obtained by either freezing the PLL or by using a very slow one [17]. However, this type of solution will not be able to provide efficient harmonic compensation. As such, a trade-off between the dynamic response and smooth transient behavior has to be made in PLL parameter selection.

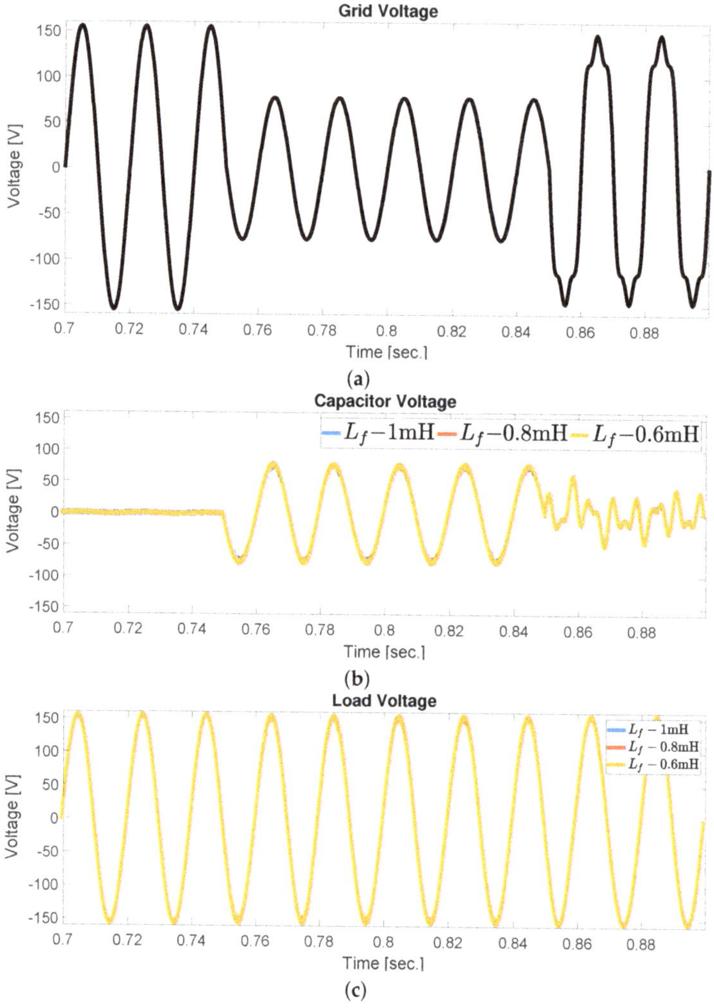

Figure 12. Simulation results for controller robustness check. (**a**) Grid voltage; (**b**) capacitor voltage; (**c**) load voltage.

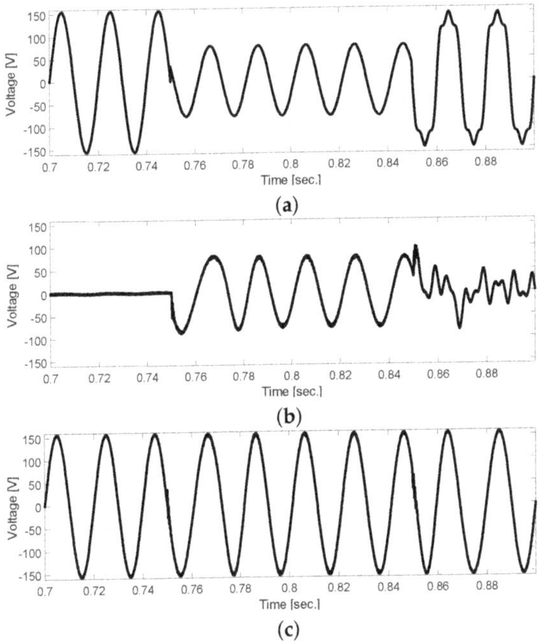

Figure 13. Simulation results for combined voltage sag and phase change at the grid. (**a**) Grid voltage; (**b**) capacitor voltage; (**c**) load voltage.

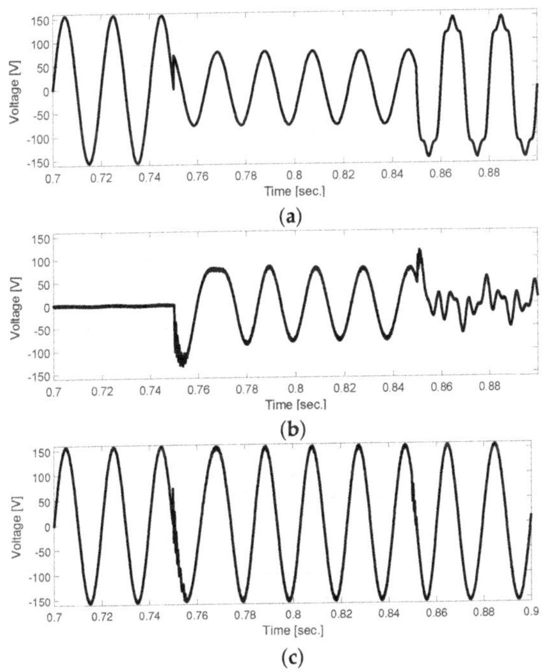

Figure 14. Simulation results for combined voltage sag, frequency, and phase change at the grid. (**a**) Grid voltage; (**b**) capacitor voltage; (**c**) load voltage.

6. Conclusions

In this paper, an enhanced single-phase quasi type-1 PLL was proposed to generate the reference compensation voltage for the multi-functional operation of a single-phase dynamic voltage restorer. A super-twisting sliding-mode controller was also proposed to track the reference voltage. The developed PLL is highly robust to grid voltage harmonics, which resulted in very low total harmonic distortion at the sensitive load-side. Moreover, thanks to a super-twisting controller, fast tracking of the compensation voltage was also achieved. Stability analysis and tuning of the PLL are presented by using small-signal modeling. The developed control method has been validated in a laboratory-scale prototype. The experimental results show that the proposed controller is very effective in compensating any grid voltage abnormalities, which in turn contributes to keeping the voltage at the sensitive load-side to remain very close to the ideal reference voltage.

Author Contributions: Conceptualization, H.A., S.B., H.K. and M.B.; methodology, H.A.; software, H.A. and S.B.; validation, S.B.; writing—original draft preparation, H.A.; writing—review and editing, H.A., S.B., H.K. and M.B. All authors have read and agreed to the published version of the manuscript.

Funding: H. Ahmed is funded through the Sêr Cymru programme by Welsh European Funding Office (WEFO) under the European Regional Development Fund (ERDF).

Institutional Review Board Statement: Not applicable.

Informed Consent Statement: Not applicable.

Conflicts of Interest: The authors declare no conflict of interest.

References

1. Moghassemi, A.; Padmanaban, S. Dynamic voltage restorer (DVR): A comprehensive review of topologies, power converters, control methods, and modified configurations. *Energies* **2020**, *13*, 4152. [CrossRef]
2. Yáñez-Campos, S.C.; Cerda-Villafaña, G.; Lozano-García, J.M. A Two-Grid Interline Dynamic Voltage Restorer Based on Two Three-Phase Input Matrix Converters. *Appl. Sci.* **2021**, *11*, 561. [CrossRef]
3. Abas, N.; Dilshad, S.; Khalid, A.; Saleem, M.S.; Khan, N. Power quality improvement using dynamic voltage restorer. *IEEE Access* **2020**, *8*, 164325–164339. [CrossRef]
4. Toumi, T.; Allali, A.; Abdelkhalek, O.; Abdelkader, A.B.; Meftouhi, A.; Soumeur, M.A. PV Integrated single-phase dynamic voltage restorer for sag voltage, voltage fluctuations and harmonics compensation. *Int. J. Power Electron. Drive Syst.* **2020**, *11*, 547. [CrossRef]
5. *IEEE Standard 1346-1998*; IEEE Recommended Practice for Evaluating Electric Power System Compatibility with Electronic Process Equipment. IEEE: Piscataway Township, NJ, USA, 1998. [CrossRef]
6. Lucas, A. Single-phase PV power injection limit due to voltage unbalances applied to an urban reference network using real-time simulation. *Appl. Sci.* **2018**, *8*, 1333. [CrossRef]
7. Trabelsi, M.; Komurcugil, H.; Refaat, S.S.; Abu-Rub, H. Model predictive control of packed U cells based transformerless single-phase dynamic voltage restorer. In Proceedings of the 2018 IEEE International Conference on Industrial Technology (ICIT), Lyon, France, 20–22 February 2018; pp. 1926–1931.
8. Kumar, C.; Mishra, M.K. Predictive voltage control of transformerless dynamic voltage restorer. *IEEE Trans. Ind. Electron.* **2015**, *62*, 2693–2697. [CrossRef]
9. Ahmed, H.; Salgado, I.; Chairez, I.; Benbouzid, M. Robust gradient estimator for unknown frequency estimation in noisy environment: Application to grid-synchronization. *IEEE Access* **2020**, *8*, 70693–70702. [CrossRef]
10. Ahmed, H.; Biricik, S.; Benbouzid, M. Enhanced Frequency Adaptive Demodulation Technique For Grid-Connected Converters. *IEEE Trans. Ind. Electron.* **2021**, *68*, 11053–11062. [CrossRef]
11. Remya, V.; Parthiban, P.; Ansal, V.; Nandakumar, A. Single-phase DVR with semi-Z-source inverter for power distribution network. *Arab. J. Sci. Eng.* **2018**, *43*, 3135–3149. [CrossRef]
12. Satputaley, R.J.; Borghate, V.B. Performance analysis of DVR using "new reduced component" multilevel inverter. *Int. Trans. Electr. Energy Syst.* **2017**, *27*, e2288. [CrossRef]
13. Ahmed, H.; Biricik, S.; Benbouzid, M. Linear Kalman Filter-Based Grid Synchronization Technique: An Alternative Implementation. *IEEE Trans. Ind. Inform.* **2021**, *17*, 3847–3856. [CrossRef]
14. Ahmed, H.; Benbouzid, M. Simplified Second-Order Generalized Integrator - Frequency-Locked Loop. *Adv. Electr. Electron. Eng.* **2019**, *17*, 405–412. [CrossRef]
15. Ahmed, H.; Pay, M.L.; Benbouzid, M.; Amirat, Y.; Elbouchikhi, E. Hybrid estimator-based harmonic robust grid synchronization technique. *Electr. Power Syst. Res.* **2019**, *177*, 106013. [CrossRef]

16. Meena, A.; Islam, S.; Anand, S.; Sonawane, Y.; Tungare, S. Design and control of single-phase dynamic voltage restorer. *Sādhanā* **2017**, *42*, 1363–1375. [CrossRef]
17. Nielsen, J.G.; Newman, M.; Nielsen, H.; Blaabjerg, F. Control and testing of a dynamic voltage restorer (DVR) at medium voltage level. *IEEE Trans. Power Electron.* **2004**, *19*, 806–813. [CrossRef]
18. Biricik, S.; Komurcugil, H. Optimized sliding mode control to maximize existence region for single-phase dynamic voltage restorers. *IEEE Trans. Ind. Inform.* **2016**, *12*, 1486–1497. [CrossRef]
19. Biricik, S.; Khadem, S.K.; Redif, S.; Basu, M. Voltage distortion mitigation in a distributed generation-integrated weak utility network via a self-tuning filter-based dynamic voltage restorer. *Electr. Eng.* **2018**, *100*, 1857–1867. [CrossRef]
20. Biricik, S.; Redif, S.; Özerdem, Ö.C.; Khadem, S.K.; Basu, M. Real-time control of shunt active power filter under distorted grid voltage and unbalanced load condition using self-tuning filter. *IET Power Electron.* **2014**, *7*, 1895–1905. [CrossRef]
21. Rajkumar, K.; Parthiban, P. Grid voltage detection method based on a novel adaptive notch filter for control of transformerless dynamic voltage restorer. In Proceedings of the 2020 IEEE First International Conference on Smart Technologies for Power, Energy and Control (STPEC), Nagpur, India, 25–26 September 2020; pp. 1–6.
22. Mellouli, M.S.; Hamouda, M.; Slama, J.B.H.; Al Haddad, K. A Third-order MAF Based QT1-PLL That is Robust against Harmonically Distorted Grid Voltage with Frequency Deviation. *IEEE Trans. Energy Convers.* **2021**, *36*, 1600–1613. [CrossRef]
23. Golestan, S.; Freijedo, F.D.; Vidal, A.; Guerrero, J.M.; Doval-Gandoy, J. A quasi-type-1 phase-locked loop structure. *IEEE Trans. Power Electron.* **2014**, *29*, 6264–6270. [CrossRef]
24. Ahmed, H.; Tir, Z.; Verma, A.K.; Ben Elghali, S.; Benbouzid, M. Quasi Type-1 PLL With Tunable Phase Detector for Unbalanced and Distorted Three-Phase Grid. *IEEE Trans. Energy Convers.* **2021**. [CrossRef]
25. Wu, C.; Xiong, X.; Taul, M.G.; Blaabjerg, F. On the Equilibrium Points in Three-Phase PLL Based on the d-axis Voltage Normalization. *IEEE Trans. Power Electron.* **2021**, *36*, 12146–12150. [CrossRef]
26. Gautam, S.; Xiao, W.; Lu, D.D.C.; Ahmed, H.; Guerrero, J.M. Development of Frequency-Fixed All-Pass Filter based Single-Phase Phase-Locked Loop. *IEEE J. Emerg. Sel. Top. Power Electron.* **2021**. [CrossRef]
27. Ahmed, H.; Benbouzid, M. On the enhancement of generalized integrator-based adaptive filter dynamic tuning range. *IEEE Trans. Instrum. Meas.* **2020**, *69*, 7449–7457. [CrossRef]
28. Golestan, S.; Guerrero, J.M.; Vasquez, J.C. Single-Phase PLLs: A Review of Recent Advances. *IEEE Trans. Power Electron.* **2017**, *32*, 9013–9030. [CrossRef]
29. Omar, A.I.; Aleem, S.H.A.; El-Zahab, E.E.; Algablawy, M.; Ali, Z.M. An improved approach for robust control of dynamic voltage restorer and power quality enhancement using grasshopper optimization algorithm. *ISA Trans.* **2019**, *95*, 110–129. [CrossRef]
30. Li, Y.W.; Blaabjerg, F.; Vilathgamuwa, D.M.; Loh, P.C. Design and comparison of high performance stationary-frame controllers for DVR implementation. *IEEE Trans. Power Electron.* **2007**, *22*, 602–612. [CrossRef]
31. Komurcugil, H.; Biricik, S.; Bayhan, S.; Zhang, Z. Sliding Mode Control: Overview of Its Applications in Power Converters. *IEEE Ind. Electron. Mag.* **2021**, *15*, 40–49. [CrossRef]
32. Biricik, S.; Komurcugil, H.; Ahmed, H.; Babaei, E. Super twisting sliding mode control of DVR with frequency-adaptive Brockett oscillator. *IEEE Trans. Ind. Electron.* **2021**, *68*, 10730–10739. [CrossRef]
33. Biricik, S.; Komurcugil, H.; Tuyen, N.D.; Basu, M. Protection of Sensitive Loads Using Sliding Mode Controlled Three-Phase DVR With Adaptive Notch Filter. *IEEE Trans. Ind. Electron.* **2019**, *66*, 5465–5475. [CrossRef]
34. Chedjara, Z.; Massoum, A.; Wira, P.; Safa, A.; Gouichiche, A. A fast and robust reference current generation algorithm for three-phase shunt active power filter. *Int. J. Power Electron. Drive Syst.* **2021**, *12*, 121–129. [CrossRef]
35. Sevilmiş, F.; Karaca, H. A fast hybrid PLL with an adaptive all-pass filter under abnormal grid conditions. *Electr. Power Syst. Res.* **2020**, *184*, 106303. [CrossRef]
36. Ramezani, M.; Golestan, S.; Li, S. Non-frequency sensitive all-pass filter based single-phase PLLs. In Proceedings of the 2016 IEEE/PES Transmission and Distribution Conference and Exposition (T&D), Dallas, TX, USA, 3–5 May 2016; pp. 1–5.
37. Levant, A. Higher-order sliding modes, differentiation and output-feedback control. *Int. J. Control* **2003**, *76*, 924–941. [CrossRef]
38. Derafa, L.; Benallegue, A.; Fridman, L. Super twisting control algorithm for the attitude tracking of a four rotors UAV. *J. Frankl. Inst.* **2012**, *349*, 685–699. [CrossRef]

MDPI
St. Alban-Anlage 66
4052 Basel
Switzerland
Tel. +41 61 683 77 34
Fax +41 61 302 89 18
www.mdpi.com

Applied Sciences Editorial Office
E-mail: applsci@mdpi.com
www.mdpi.com/journal/applsci

www.ingramcontent.com/pod-product-compliance
Lightning Source LLC
Chambersburg PA
CBHW061610210926
51850CB00009B/2279
9783036556512